URBAN PLANNING and DEVELOPMENT APPLICATIONS of
GIS

Sponsored by
Geographic Information Systems Committee of the Urban
    Planning and Development Division
American Society of Civil Engineers

Edited by

Said Easa and Yupo Chan

 **American Society
of Civil Engineers**

1801 ALEXANDER BELL DRIVE
RESTON, VIRGINIA 20191–4400

Abstract: This book presents the state-of-the-art in urban planning and development applications of geographic information systems (GIS). The book consists of two sections: GIS methodologies and GIS applications. The methodologies section is intended to provide the reader with GIS concepts and techniques, including GIS technology implementation, remote sensing, trends in spatial databases, and linear referencing systems. The applications section is intended to illustrate the capabilities of GIS and related technologies in urban planning and development. The applications cover a wide range of topics, including regional planning, transportation, public utilities, stormwater and waste management, cultural and resources management, environmental assessment, program evaluation and policy analysis, and education. These methodologies and applications have great potential to assist practicing professionals with new ideas to solve the continually challenging problems in planning and development. The book concludes with some observations on technology and database issues facing the profession.

Library of Congress Cataloging-in-Publication Data

Urban planning and development applications of GIS / edited by Said Easa and Yupo Chan.
    p.   cm.
   "Sponsored by Geographic Information Systems Committee of the Urban
Planning and Development Division, American Society of Civil Engineers."
   Includes bibliographical references and index.
   ISBN 0-7844-0461-5
   1. Information storage and retrieval systems--City planning. 2. Geographic information systems. I. Easa, Said. II. Chan, Yupo. III. American Society of Civil Engineers. Geographic Information Systems Committee.

HT166 .U7385 1999
711'.4'0285--dc21   99-054569

PREFACE

Urban Planning and Development Applications of GIS is intended to present state-of-the-art applications of geographic information systems to help both beginners and specialists better understand current and emerging developments in the field. The idea for this book arose from the need to provide the urban planning and development community with an integrated document on current GIS applications that sporadically appear in the literature. We hope that the book will spark interest in the reader to employ the presented concepts and ideas in his/her own endeavor and to expand the array of GIS applications.

The book is organized into two sections: GIS methodologies and GIS applications. The methodologies present GIS basic concepts and new developments. The methodologies are presented in four chapters: GIS technology and implementation, remote sensing, trends in spatial databases, and implementation of linear referencing systems in GIS. We included separate chapters on the last three topics because of the important role they play in GIS. The applications were selected so that they would cover all major areas of urban planning and development: regional planning, transportation, public utilities, stormwater and waste management, cultural and resources management, environmental assessment, socioeconomic development, and academic education. In each of these nine areas we have presented one chapter that either reviews the state-of-the-art in that area or presents a case study that illustrates the enormous capabilities of GIS. The book also includes a guide to the many information sources available in the Internet (Appendix 1) and a list of acronyms (Appendix 2).

A glimpse of some special features covered in the book would be helpful. Important advances in GIS such as spatial decision-support systems and GIS integration with analytical models (like expert systems, spatial statistics, and optimization) are discussed. Three-dimensional analysis, an emerging development in GIS, is nicely illustrated in a number of applications, including environmental assessment and cultural/natural resources. Incorporating real-time information into GIS for public utilities and stormwater/waste management is echoed in the book. The Voronoi diagram, one of the newest approaches in planning, is used in an emergency-response application. In another application on socioeconomic development, GIS allowed traditional econometric models to include a spatial dimension.

We wish to express our thanks to the executive committee of the ASCE Urban Planning and Development Division, in particular Kurt Ardaman and Larry Mugler for their support during the development of this book. Special mention should be made of the ASCE staff, Charlotte McNaughton and Susan Fisher for their guidance and diligent work during the book production. The help of Songnian Li and M.T. Herzog in preparing Appendix 1 is gratefully acknowledged. Our sincere appreciation goes to the authors for their enthusiasm and contributions that have enriched the field of GIS.

Said Easa, Lakehead University
Yupo Chan, Air Force Institute of Technology

THE EDITORS

Said Easa is currently Professor in the Department of Civil Engineering, Lakehead University, Thunder Bay, Ontario. He earned his M.Eng. from McMaster University and Ph.D. from University of California at Berkeley in 1982. He has conducted research in a wide range of transportation engineering and planning areas, and published over 130 refereed journal articles and was editor of seven books. Since 1995, Easa has chaired the Geographic Information Systems Committee of the Urban Planning and Development Division, American Society of Civil Engineers (ASCE).

Easa's professional work included serving as Chair of the Executive Committee of the Urban Transportation Division of ASCE and that of CSCE (Canadian Society for Civil Engineering). He is a member of CSCE Board of Directors and Vice-Chair of the General Administration Committee. In 1996, he initiated and chaired the national series of CSCE transportation conferences, and co-chaired the ASCE conference on "*Transportation, Land Use, and Air Quality: Making the Connection,*" Portland, Oregon, 1998. Easa led a delegation to China in 1998 focusing on sustainable transportation systems and land-use planning. His work received several national best paper awards, including the 1997 CSCE Keefer Medal and its 1998 Honorable Mention. He also received the Lakehead University 1993 Distinguished Researcher award and 1998 Contributions to Research award.

Yupo Chan is currently Professor in the Department of Operational Sciences, Air Force Institute of Technology, Wright-Patterson AFB Ohio. He earned all his degrees (Bachelor, Masters, and Ph.D.) from the Massachusetts Institute of Technology. Chan has 27 years of postdoctoral experience in industry, universities, and government, including an honorary assignment as a Congressional Fellow. His research interests center around transportation science, facility-location and land-use, spatial-temporal information, multicriteria decision-making, and optimization.

Chan is the author of two books related to this volume: *Location Theory and Decision Analysis* (ITP/South-Western) and *Location, Transport, and Land Use: Modeling Spatial-Temporal Information* (Springer-Verlag) and editor of several books. He has been active in the American Society of Civil Engineers, including service as Chair of Management Group C, which includes four divisions: Air Transport, Highways, Urban Transportation, and Urban Planning and Development. He received the 1991 Koopman Prize of the Institute of Operations Research and Management Sciences. Chan swims seven days a week, which explains why his hair is bleached grayer and grayer every time you see him.

CONTRIBUTORS

William Bachman
Center for Geographic Information
 Systems
Georgia Institute of Technology
Atlanta, GA 30332
wbachman@arch.gatech.edu

Yupo Chan
Department of Operational Sciences
Air Force Institute of Technology
(AFIT/ENS)
2950 P St, Bldg. 640
Wright-Patterson AFB, Ohio 45433
ychan@alum.MIT.edu

David J. Coleman
Department of Geodesy and
 Geomatics Engineering
University of New Brunswick
Frederickton, NB
Canada E3B 5A3
dcoleman@unb.ca

Kristina Dalton
P.O. Box 19161
Tucson, AZ 85731
dalton@theriver.com

Lisa DeLorenzo
School of Public Affairs
Arizona State University
Tempe, AZ 85287
lisa.delorenzo@asu.edu

Said Easa
Department of Civil Engineering
Lakehead University
Thunder Bay, ON
Canada P7B 5E1
seasa@lakeheadu.ca

Shauna Hallmark
School of Civil and Environmental
 Engineering
Goergia Institute of Technology
Atlanta, GA 30332
shallmar@trec.ce.gatech.edu

Elizabeth A. Harper
Private Consultant
19 Shepard St., Unit 31
Cambridge, MA 02138
harper@appgeo.com

M.T. Herzog
Foothill Engineering Consultants,
 Inc.
350 Indiana St., Suite 415
Golden, CO 80401
fec@rmi.net

K. Jeyapalan
Department of Civil and
 Construction Engineering
394 Town Engineering Bldg.
Iowa State University
Ames, IA 50011
Jeyapalan@ccelab.iastate.edu

John W. Labadie
Department of Civil Engineering
Colorado State University
Ft. Collins, CO 80523
labadie@engr.colostate.edu

Songnian Li
Department of Geodesy and
 Geomatics Engineering
University of New Brunswick
Frederickton, NB
Canada E3B 5A3
k0b01@unb.ca

Larry G. Mugler
Development Services
Denver Regional Council of
Governments
2480 W. 26th Ave., #200B
Denver, CO 80211
lmugler@aol.com

Shashi Sathisan Nambisan
Dept of Civil & Environmental
  Engineering
UNLV Transportation Research
  Center
4505 Maryland Parkway,
Box 454015
Las Vegas, NV 89154
shashi@ce.unlv.edu

Wende O'Neill
Westat Inc.
1650 Research Blvd.
Rockville, MD 20850
(301) 610-4816
oneillw@westat.com

Emelinda M. Parentela
Civil Engineering Department
California State University Long
Beach, VEC104
Long Beach, CA 90840
parent@engr.csulb.edu

Terence T. Quinn
Denver Planning Office
200 W. 14th Ave., #203
Denver, CO 80204
quinntt@ci.den.co.us

Siva Ravada
Spatial Products Division
Oracle Corporation
One Oracle Drive
Nashua, NH 03062
sravada@us.oracle.com

Wayne Sarasua
Department of Civil Engineering
Clemson University
Clemson, SC 29634
sarasua@clemson.edu

Jayant Sharma
Spatial Products Division
Oracle Corporation
One Oracle Drive
Nashua, NH 03062
jsharma@us.oracle.com

Yaoyu Shi
Department of Civil Engineering
Ningxia University
Yingchuan, Ningxia 750021
P.R. China
tugc@nxu.edu.ca

Reginald R. Souleyrette
Center for Transportation Research
  and Education, and
Department of Civil and
  Construction Engineering
Iowa State University
Ames, IA 50011
reg@iastate.edu

Tim R. Strauss
Department of Geography
University of Northern Iowa
Sabin 5
Cedar Falls, IA 50614
tim.strauss@uni.edu

CONTENTS

Section II: GIS Applications

CHAPTER 1

Introduction

Said Easa and Yupo Chan

The purpose of this book is to present current applications of geographic information systems (GIS) in various fields of urban planning and development. The advances in GIS technology in recent years have been impressive, and so was the wide range of its applications. The fields of urban planning and development applications involve the planning, design, operation, and management of infrastructure and resources. Several innovative GIS technologies have been used in these applications, including three-dimensional (3-D) analysis, integration with analytical models (like expert systems, statistical methods, and optimization techniques), and distributed databases. In addition, extensive experience has been gained through a large number of applications in these fields, where GIS has proven to be a powerful tool for decision making.

The applications presented in this book include a combination of state-of-the art reviews and specific case studies that illustrate the spectacular capabilities of GIS. This variety of applications is intended to provide insight into current status of GIS use in practice and to help the reader effectively apply GIS tools to his/her own practical endeavor. Indeed, the wide variety of problems and solutions presented can be easily adapted to solving many similar problems in planning and development.

The book is organized into two sections: GIS methodologies and GIS applications. The section on methodologies is intended to provide the reader with a basic understanding of GIS concepts. The section covers four basic topics: GIS technology and implementation, remote sensing, trends in spatial databases, and linear referencing systems. While the chapter on GIS technology and implementation covers all aspects of GIS, we felt the need to present separate chapters on remote sensing and spatial databases (information rarely covered in GIS books) because of the key roles these technologies play in geographic information systems. Linear referencing systems were also treated separately because they are quite useful in special GIS applications, especially transportation.

The section on applications was designed to cover all major application fields in urban planning and development. These fields are regional planning, infrastructure management (transportation, public utilities, and stormwater/waste), resource management, environmental assessment, socioeconomic development, emergency management, and education. In each of these nine fields, we included a chapter that reviews the current state-of-the-art in the field or presents an innovative case study. The way of organizing this book into two sections on methodologies and applications has been rewarding in our opinion. The contributing authors have responded beautifully to this taxonomy.

GIS METHODOLOGIES

The organization of the book chapters is shown in Table 1. In Chapter 2, *GIS Technology and Implementation,* Herzog presents a comprehensive overview of GIS technology. She introduces many key technology concepts, from spatial data sources, structures, and metadata (data about data) through functionality and GIS integration. In particular, she discusses GIS integration with other tools such as analytical models (expert systems and mathematical optimization) and spatial decision-support systems, along with practical examples. GIS implementation at both the organizational level and as a method to share information throughout society is also addressed. This chapter serves as a great way to open the discussions.

Following this, Jeyapalan has done a great job in surveying the field of *Remote Sensing* in Chapter 3. The topics range from remote sensing history, process (e.g. electromagnetic spectrum and atmospheric windows), devices (e.g. aerial camera and microwave sensing), approaches to data analysis (e.g. spectral response patterns and photo interpretation), to the complementary relationship between remote sensing and GIS. This chapter constitutes a 'primer' for those who want to have an understanding of remote sensing.

Ravada and Sharma in Chapter 4, *Trends in Spatial Databases,* have spoken from their vast experience on how to manage geographic information. The object-relational database management system, combining the best features of relational database and object-oriented database schemes, is proposed as the way to overcome the difficulties faced by existing database management systems. The authors identify some of the research issues and recent accomplishments in spatial-database systems, such as spatial-data types and operators, spatial-query languages and processing techniques, spatial indexing, and clustering techniques.

O'Neill and Harper in Chapter 5, *Implementation of Linear Referencing Systems in GIS,* have pointed out the relationship between linear referencing system (LRS) and GIS for transportation. They present an overview of LRS with a discussion of the advantages and disadvantages of various linear referencing methods. The discussion is both instructional and forward looking. GIS can be viewed, like anything else, in a couple of different angles. One view is that all GIS data must be established within a coordinate system that can serve a spatial reference system. The United States Geological Survey (USGS) maps have traditionally been serving as a locational reference, until global positioning systems (GPS) gradually provided accurate survey-coordinate information with time and cost savings. LRS, on the other hand, is a special GIS that builds upon a "one-dimensional" referencing scheme.

GIS APPLICATIONS

Turning our attention to applications, in Chapter 6, *Regional Planning (Activity-Allocation Modeling),* Mugler and Quinn discuss the way one regional planning agency, the Denver Regional Council of Governments, has utilized its

Table 1 GIS Methodologies and Applications in Urban Planning and Development

Section	Area	Chapter number
Methodologies	Technology and implementation	2
	Remote sensing	3
	Spatial databases	4
	Linear referencing systems	5
Applications	Regional planning	6
	Infrastructure management	
	Transportation	7
	Public utilities	8
	Stormwater and waste	9
	Resource management	10
	Emergency management	11
	Environmental assessment	12
	Socioeconomic development	13
	Education	14

GIS to assist in the preparation of demographic forecasts (such as wastewater flows, traffic conditions on major roadways or transit lines, and the number of elderly needing public services). The authors point out that demand data are usually aggregated according to some arbitrary spatial unit, such as a census tract or a traffic analysis zone. GIS is one way to define such a common geographic unit. Conversion from existing databases to such a common format, however, needs to be handled with care. When executed in the correct way, we observe that remote sensing and GIS have the potential to describe land-use not only in terms of two-dimensional geographic and socioeconomic information, but also to include elevation (or time) as a third dimension. This was also pointed out in other chapters by Ravada and Sharma, Dalton, and others.

Souleyrette and Strauss in Chapter 7, *Transportation,* did a comprehensive survey of various areas of transportation-planning and traffic-engineering applications. Let us cite an example. Floating cars are traditionally used to collect travel-time and delay information. Instead of manual labor, GPS receivers can be used to log time and position points of a floating car. Once collected, the GPS data-files are corrected for selective-signal availability and/or ionospheric distortion using publicly available base-station correction files. The data points are then linked spatially to a previously prepared map-layer of numbered roadway segments and buffers. This step adds a segment identifier to each data-point in the corrected GPS data-file. The edited data-point files are now converted to a data-based format readable by a standard statistical software-package, which is subsequently used for reduction and data processing. The graphic displays afforded by GIS of travel-time and delay data are also a highly desirable feature.

In Chapter 8, *Public Utilities,* Li et al. review the automated mapping/facility management (AM/FM) systems. They examine current AM/FM applications and implementation issues in public utilities and explore the relationship between AM/FM systems and some emerging technologies, such as Internet-based technology, GIS toolkits, and object-oriented technology. The authors also echo the need for real-time information and, as an example, describe the supervisory control and data acquisition (SCADA) system.

Herzog and Labadie's Chapter 9, *Stormwater and Waste Management,* like other chapters, adds to the richness of the book. The authors did a comprehensive job surveying stormwater/wastewater, integrated solid-waste management, and hazard-waste management. The tie-in to multi-criteria decision-analysis is to be applauded, in view of the need to respond to multiple stakeholders with different aspirations. Echoing a common observation reported elsewhere in this book, the authors discuss real-time applications of GIS. In the same vein, the authors insightfully point out three key features of GIS: overlay, buffers, and map-ematics. *Overlays* combine attributes from different map layers, *buffers* analyze the proximity of one object to another, and *map-ematics* contain the analytical tools for spatial modeling.

Dalton in Chapter 10, *Cultural and Natural Resources,* describes a few GIS applications in cultural resources in the field of archaeology (such as predictive modeling, site risk analysis, and line-of-sight analysis) and in natural resources (such as vegetation studies, biological evaluations, and mine planning). Dalton has shown how GIS can offer innovative solutions to problems that were previously solved by extensive field work or manual mapping. She also echoed others in the emerging use of 3-D modeling (Eichelberger 1998).

Parenttela and Sathisan-Nambisan in Chapter 11, *Emergency Response (Disaster Management),* refer to some inherent analytical features in most geographic information systems, including straightforward functions such as 'address coding' and more sophisticated features such as Thiessen-Polygon (Voronoi-diagram) representation. We would like to interject that the use of Voronoi diagram is one of the newest approaches in urban planning and development. Chan (2000) has shown that many spatial routines can be easily implemented in Voronoi diagrams, while it is much less straightforward in a regular GIS data-structure. The authors illustrated these features nicely with a case study from Yucca Mountain, Nevada, complete with very convincing displays. Following the location/allocation discussions above, they were able to show the locations of fire stations, hazardous-materials stations, and police stations. Most important, the response time from these service stations to an emergency can be well predicted from GIS.

Sarasua et al. in Chapter 12, *Environmental Assessment of Transportation-Related Air Quality,* present a rich study replete with GIS tools, ranging from data input to modeling. The GIS integration hierarchy is particularly helpful in understanding the role of GIS in problem solving. One problem faced by environmental planners is the location of obnoxious facilities, such as smoke stacks and landfills. Organized around a GIS, the general location problem boils down to a processing of these data files according to our observations:

- *dataset/coverage*: the name of the dataset that will be used in the location/allocation process, consisting of points, polygons, lines (street network) with a specified coordinate system and ancillary attributes
- *demand locations*: a set of locations that need to be serviced, each of which may have a weight associated with them (such as population)
- *candidate locations*: a set of potential locations where the site could be located
- *distances*: a method to compute the distances or travel costs, whether Euclidean or shortest-path
- *models*: the method in which the locations will be chosen
- *sites*: the number of actual sites that need to be located.

DeLorenzo in Chapter 13, *Program Evaluation and Policy Analysis,* introduces important tools that are not necessarily familiar to traditional civil engineers. She takes on the case study of residential mobility and downtown redevelopment to illustrate these tools. GIS allows traditional econometric methods to be broadened to include the spatial dimension. In our perspective, spatial econometric models, when viewed in a broader context, can be described as special cases of spatial time-series, or geographic time-series. Perhaps the most common form is the spatial autoregressive-model, which establishes the correlation between neighboring geographic-units. The linkage lies in the way the 'spatial' weight matrix is defined. Here, the weight matrix quantifies the contiguity relationship between economic sectors. In other words, how does the residential sector interact with the employment sector, and how does it interact with the retail sector, and so on? It can be seen therefore that the concept of spatial econometrics is rather far reaching. It is best viewed as a prominent (yet specialized) branch of statistics that can delineate, among other things, redevelopment policies in many of our deteriorating downtowns.

In Chapter 14, *Civil Engineering Education,* Easa et al. review current status of GIS uses and developments in civil engineering education and present guidelines regarding GIS course setting, education methods, and infrastructure needs. They surveyed sample universities to gain insight toward the extent and fashion that GIS has been and could be included in educational programs. Meanwhile, debate is going on regarding whether land-base data exhibited in GIS should be created, prepared, and modified under the supervision of a licensed land surveyor or professional engineer authorized to practice land surveying (Geo Info Systems 1998). Irrespective of its outcome, such debate would certainly have profound bearing upon the training and responsibility of future civil engineers in our opinion.

These sections and chapters reflect the real beauty of GIS: its ability to support quality decision-making for complex problems in all fields of urban planning and development. In addition, the book provides lessons learned from practical experience in using GIS by both industry and academic professionals. It discusses traditional and emerging GIS concepts and their potential applications in a coherent blend of chapters on state-of-the-art reviews and case studies. It alerts the reader, whenever applicable, to opportunities in which emerging technologies can be successfully applied. We hope that the book will spark the interest of both beginners and specialists and help expand the array of GIS applications in urban planning and development.

REFERENCES

Chan, Y. (2000). *Location, transport, and land-use: modeling spatial-temporal information*. Springer-Verlag, New York, N.Y.

Eichelberger, P. (1998). "3D GIS: the necessary next wave" *Geo Info Systems*. October.

Geo Info Systems. (1998). "GIS vision: industry professionals share." September.

Section 1

GIS Methodologies

CHAPTER 2

GIS Technology and Implementation

M. T. Herzog

A better understanding of the basics and implementation issues of Geographic Information Systems (GIS) is essential to their successful application, especially to complex problems. GIS itself is quite a complex topic, and is more so today as its functionality and uses continue to expand. This chapter introduces many key GIS technology concepts, from spatial data sources and structures through functionality and methods to integrate GIS with other tools. GIS implementation at both the organizational level and as a method to share information throughout society is also addressed. The implementation information may be particularly helpful to managers, as GIS usually requires the implementing organization to manage and utilize data in an entirely new way.

INTRODUCTION

Nearly three decades since Geographic Information Systems (GIS) began to develop into the form we know them today, an adequate definition of GIS remains difficult. Although somewhat simplistic, one way to put GIS in context is to think of the acronym to mean geographic information (*integrating and generating*) systems.

Computer-based GIS *integrates* data from diverse disciplines and various formats to *generate* useful information about an area of the Earth at an appropriate scale. This is accomplished by using the unique tools and methods provided in GIS software packages for capturing, organizing, processing, and analyzing spatially referenced data. Any data that can be mapped, can become an integral part of GIS by relating the geographical features (e.g. city parcels) to a linked database of attributes (e.g. owner and property value). GIS does not necessarily have to link the database attributes of mapped features to a location on the Earth, but to any spatially referenced system. For example, future GIS technology may advance our understanding of spatial patterns in the human genome or outer space.

GIS is often confused with Automated-Mapping because both can produce mapped output. However, GIS integrates data quite differently from mapping systems. Instead of placing all geographical features on a single layer, GIS is usually constructed to place related geographical features into separate, *thematic* layers, and each layer is associated with its own table of attributes in a database management system (DBMS). GIS can then relate the thematic information in the database tables based on mapped feature position (e.g. to spatially analyze elevation, wildlife, and vegetation layers together). Additionally, since GIS stores mapped feature positions in a mathematical way (called *topology*), each feature can be related to other features within a single thematic layer by relationships of proximity and adjacency among others. This *geo-relational* model of data structure (as well as the more-advanced spatial data models being developed) permits a wide array of spatial operations to be performed on data in GIS.

The powerful information-generating aspect of GIS can be better understood by considering how humans think. Research indicates that people have trouble remembering more than seven pieces of data at once in short-term memory; however, we can absorb almost limitless information instantly once it has been arranged into a recognizable pattern (Gore 1998). Therefore, by allowing vast stores of data to be integrated and analyzed in a wide variety of ways, GIS permits us to better understand problems and more-effectively investigate solution scenarios. Usually, spatial analysis results are mapped in GIS, the way the brain may best sum up the import of the newly synthesized information. GIS also incorporates a variety of other presentation methods, including reports, statistical summaries, three-dimensional (3-D) renderings, *hot links*, and animation. Through a hot link, when a feature such as a building footprint is selected, a pop-up window can display a drawing of its interior spaces, a photo of its facade, a database of tenants, a scanned site plan, a video of its history, subdivision documents, and/or a recorded advertisement. Methods of data integration in GIS are expanding and being tailored for specific purposes to eventually allow all data in any format to be organized into a single digital framework.

Although at first GIS were the domain of expert users, today's graphical user interfaces (GUI) for GIS are intuitive, interactive, and allow for easy navigation through a variety of sophisticated routines. In this way, even complex GIS operations have been simplified. The development of Intelligent Multimedia Presentation Systems (IMMPS) will probably allow future GIS users to work even more interactively with GIS software, and perhaps even ask useful, but unexpected questions instead of rigid queries (Galetto and Spalla 1996).

Improved efficiency and the power of visualization are two of the key reasons why many organizations that traditionally use DBMS or other information management systems (IMS) have switched to GIS. Another reason is that many management and problem-solving tasks require referencing both tabular and mapped data, which GIS best integrates. An additional benefit is that, in the process of building GIS, major errors in both geographical features and database attributes may become more apparent, allowing the accuracy and precision of system input to be improved in the process. Also, the popularity of GIS stems from its ability to integrate with almost any other software through available task and discipline-specific applications. GIS functionality is also being embedded directly into other software programs to extend their usefulness. Finally, the measure of effectiveness of an organization by both its clients and peers is linked to how well it manages resources, especially the productivity of its personnel, which a well-implemented GIS can help optimize.

SPATIAL DATA SOURCES

Traditional Data Collection Methods

In the past, most digital spatial datasets were developed from paper maps or mylars that were fastened to a digitizing tablet, that was then calibrated to real world coordinates and units, and lines on the map were traced with a digitizer to generate a digital duplicate. The digital drawing was then converted into a vector layer in GIS and joined to tabular attributes of the mapped features to create a *spatial dataset* or *coverage*. Although today this procedure is still practiced to some extent, it is now common to scan and vectorize the maps, and then fix errors using tailored editing tools within GIS.

Aerial photography is also an indispensable means for obtaining updated base information (e.g. road centerlines and building footprints). The photos are taken using digital cameras from small aircraft, so that they can be directly processed with software to remove distortion resulting from topographic relief, airplane tilt, and other sources, in the process of *ortho-rectification*. Surveyed ground control points are used to register photos to real world coordinates. Usually, either software is used to extract vector features from the photos or *heads-up* digitizing is used to trace the features directly from the photo on-screen. The ortho-rectified photo itself often serves as a base layer in GIS to assist in locating features on the landscape. Using *photogrammetry*, rough topographic maps of vector elevation contours or raster Digital Elevation Models (DEM) can also be produced from aerial photos and limited ground control survey data.

For some time, surveyed data have been collected and processed in a digital format, so the results can be directly used to generate mapped features for GIS. Many GIS also include Coordinate Geometry (COGO) procedures for transforming survey data into graphic format.

Global Positioning System

An important technology that has simplified surveying is the Global Positioning System (GPS). A hand-held GPS receiver captures signals from a worldwide network of GPS satellites to calculate its position on the Earth. The receiver must collect data emitted from at least four visible GPS satellites to determine its approximate distance from each of them and correct for some errors. The receiver can then perform trilateration calculations to pinpoint its latitude and longitude. Atmospheric conditions and sometimes even intentional error added by military entities can degrade the accuracy of GPS positioning by more than 100 meters.

However, differential correction methods have been developed to use data collected simultaneously from a second, stationary GPS receiver at a known position to correct errors in the data collected with a GPS receiver at an unknown position. Using even more sophisticated correction methods based on interpreting

Figure 1   The user is busy entering attribute data into the keypad as the GPS receiver records position (Courtesy of Foothill Engineering Consultants, Inc.)

patterns in the satellite signal itself, rather than just the data the signal contains, survey-grade GPS receivers can calculate elevation as well as position to within a fraction of an inch. However, due to their expense, and the fact that positional accuracy to a couple of meters is often acceptable, just differentially correcting data collected with lower quality receivers is still popular.

GPS Mapping Systems allow users to capture detailed feature and attribute information through an on-unit data dictionary, and download data directly into GIS as points, lines, and polygons (Trimble 1996). For example, in Figure 1, as the GPS receiver records position every second, the user is busily entering attribute data into a data dictionary using the keypad. Back in the office, this archaeologist differentially corrects the positions with base station data downloaded from the Internet. Then, she will process the point, line, and polygon features along with their respective attribute data directly into GIS software format in a single step.

More advanced GPS mapping systems include a pen-based or notebook computer to display a base map for easy navigation to mapping sites and better verification of feature and attribute accuracy while in the field (Trimble 1995). Combining voice recognition and GPS data collection can lead to even greater efficiency (Blaha 1998). For example, this would allow a transportation worker to describe the condition of road repair as he is driving along, while a GPS receiver with a vehicle antennae is used to record his location. Then, both types of data could be instantaneously uploaded to GIS.

Remote Sensing

Vector and raster GIS are sometimes integrated with remote sensing (RS) data for more comprehensive environmental modeling. RS technology permits data to be collected for the whole earth via sensors on satellites or for a locality by collecting data from aircraft. Sonar, thermal, optical, and radar sensors and a variety Of new types provide information about land cover, topographic relief, weather patterns, and other phenomena of interest. Recently, commercial and government organizations have been launching satellites based on newer technologies that offer better resolution of data and error-reduction (see Chapter 3).

Other Data Sources

As more and more organizations institute GIS, the cost for obtaining data from commercial vendors is falling. Many commercial vendors today offer reasonable rates even for custom spatial data requests, since the same spatial dataset can serve other clients. Many vendors are also taking advantage of the Internet to develop a wider client base. GIS users can take advantage of the Internet too not only to hunt down the best vendors and those hard-to-find themes and areas of coverage, but also to discover the many free spatial data resources available. Especially useful is the growing number of on-line listings of federal, state, and local Uniform Resource Locators (URLs) that provide free downloadable data. To find such spatial data resource lists, use a search engine with appropriate key words (like spatial, data, free, GIS, resources, and links) or go directly to the U.S. Federal Spatial Data Clearinghouse (URL: http://130.11.52.184/FGDCgateway. html). Be aware that free data can become an expensive choice though, if it does not well fit one's needs in terms of coverage, accuracy, and detail. Converting spatial data from one GIS software or exchanging format into the one needed for a user's particular GIS may also be time-consuming.

SPATIAL DATA STRUCTURES

Vector and Raster Data Models

Discrete geographical features are usually represented in a vector structure (Figure 2a) consisting of points (coded as vectors) and lines (coded as groups of points) to form chains, arcs, or polygons related by an explicit mathematical system of connectivity and order called *topology* (Krzanowski et al. 1993). Vector representations are rather compact and produce map-like output, although analysis is complex when trying to analyze thematic layers of vector data. Linear networks are often effectively expressed in a vector format, as are political jurisdictions with summary attributes.

An alternative structure usually used for continuous data such as topography is raster (grid) systems, composed of a cell matrix like a checkerboard (Figure 2b). Each cell possesses a real number value or, if the data are discrete, a set of related attributes in a linked database. Mathematical relationships between a cell and adjacent cells and between layered raster stacks are simpler than with vector models, aiding more rapid spatial analysis

However, since raster models require every unit of an area to be covered, they can require more system storage space. Presentation map quality of rasters can be

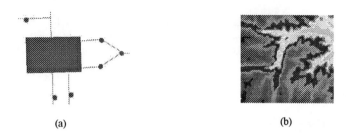

(a)   (b)

Figure 2   Data representation models: (a) vector representation of telephone poles, cables, and a sub-station on separate thematic layers and (b) raster representation of elevation contours as a digital elevation model (DEM).

poor. Describing topological relationships and projecting raster models from one coordinate system to another is also more difficult than with vector models. Since both vector and raster structures have advantages, both are used extensively. In fact, GIS often include thematic layers of both types, tools to convert between the two formats, and analysis tasks that include both.

Database Formats

A variety of database formats and structures can be used in GIS to contain the attributes associated with geographical features. Common database structures include flat file (spreadsheets), hierarchical (key-field, parent-child links), relational (tabular matrix), and object-oriented (Foote and Huebner 1996). Object-oriented databases are not used in GIS as much as relational databases yet, but they are expected to predominate soon because they allow objects rather than raster or vector primitives to reference grouped attributes directly. This makes processing and analysis tasks simpler because it allows the users to both develop and use GIS the way they think about geography every day.

For example, instead of having to make a selection of entities from a road dataset using a relational database by typing "(Road = highway, Surface = concrete, and Type = divided)", an object-oriented database would allow a user to just choose object type 'PVD' and all the paved, divided highways would be selected for further GIS processing. Furthermore, this would allow GIS processing steps to be tailored to process each object class differently, and one feature could possess multiple definitions, such as a road feature defined as both a centerline and a right-of-way area. Morphing procedures are also being developed to allow one object to be turned into another object type in such specialized systems (Maguire 1998).

Extensible DBMS

An important extension to traditional DBMS are client/server-platform databases that can carry embedded geographical feature geometry and topology in multiple formats to allow spatial datasets to be ported seamlessly between computer-aided design (CAD) systems, workstation GIS, and desktop GIS (Calkins 1997). This allows edits and changes to be completed in one system, then read and utilized in the others. Such a system can reduce redundancy and promote enterprise-wide integration to develop sophisticated GIS applications, while still allowing each department to continue to function in its own paradigm.

Triangulated Irregular Networks

Another method of representing geographical features is a Triangulated Irregular Network (TIN) that connects all points in an area with lines to form a system of triangles (Figure 3). The traditional use of TIN was to develop contour maps from point elevation data. In GIS, TIN is often used as the basis for 2.5-D renderings of topography upon which other layers of data can be draped, for example, to show a trail winding through the mountains under a setting sun. Polygons constructed by bisecting the sides of each of the triangles in a TIN at right angles are called *Thiessen* polygons. Thiessen polygons are useful for isolating the area that each individual point influences most or representing a step-like surface of attribute values to analyze patterns visually.

Advanced Structures

Two advances of GIS structures are 3-D GIS and temporal GIS. Most of today's GIS are optimized to process features with 2-D geometries. Therefore, another important extension to typical spatial data structures includes explicitly defining the third dimension of space in addition to x and y coordinates for mapping and analyzing features. 3-D GIS are typically used in geological studies, but they are also useful for modeling complex atmospheric, oceanic, and space-based phenomenon and for analyzing cityscapes.

Methods of including time as a fourth dimension in Temporal GIS (TGIS) are also being vigorously explored. This is important, because many applications of GIS could involve modeling environmental change over time or planning for development based on trends. However, including time in GIS is difficult because time is measured in different ways for different purposes (Frank 1998). Cyclical,

Figure 3 Triangular irregular network in which lines triangulate points, usually at different elevations, to appear three-dimensional upon shading.

15

ordinal, interval, hierarchical, and branching models of time could all be useful for analyzing change and exploring scenarios in GIS. Currently, time in GIS is usually handled by generating an index of feature updates in a DBMS on either an interval basis (e.g. yearly) or on an event basis (e.g. generated each time a change occurs). Loops can be used in such systems to simulate cyclical time (e.g. seasons). Using raster data or images, time can be examined by analyzing incremental changes in values between layers of data taken at different instances. Time can often be effectively visualized by animating these layers to observe geographical features as they appear, change, and disappear. TGIS can also be used to compare spatial data lagged by different intervals over a number of periods to develop predictive models, like between temperature and insect infestations.

GEOREFERENCING SYSTEMS

One of the greatest problems encountered as entities have begun to share spatial datasets is that thematic layers from different sources may lie in different locations in the coordinate space. This can occur because a different geodetic datum, unit, map projection, or reference ellipsoid was used in transforming the data from the Earth's curved surface into 2-D space in the process of *georeferencing*. Although the Earth looks relatively round from space, it is actually an imperfect ellipsoid. Therefore, mathematical models of the surface called *reference ellipsoids* are used to describe the placement of features on the Earth. Over the past 150 years, a number of manual, automated, and GPS surveys have been conducted to establish geodetic datums to help model these reference ellipsoids. Satellite snapshots of the Earth taken from a variety of angles have improved modeling accuracy in recent years.

Map Projection

Since most GIS model features are in two dimensions, and most distance and direction measurements are calculated from flat maps, usually latitude and longitude pairs are converted into planar coordinates by the *projection* method. The elevation usually must be calculated separately as a database attribute based on a standard vertical coordinate system. However, there is no way to take a curved surface and make it flat without introducing some distortion. Therefore, all projections distort one or more of the following map properties: area, shape, angles, distance, or direction. Projections are often classified according to the type of error they most successfully minimize. Small areas of the Earth are less curved than large areas, so projection errors in local area maps are usually smaller. A projection method tailored to a particular area can also be used to increase accuracy.

In the U.S., the State Plane system is popular because it uses a locally referenced datum (NAD27 or NAD83) to increase accuracy, and the projection to use for a particular state is chosen based on the shape of the state to further minimize distortion. Another common projection is the Universal Transverse Mercator (UTM) which divides the world into long strips of 6 degrees latitude each, called zones. Since UTM projections are cylindrical projections of a sphere onto a flat surface, exact scale is only true along the central meridian and along two lines equidistant from the central line. Shape and scale distortions increase as one approaches distant points at right angles from the centerline (Dana 1998). Understanding the errors that arise from different types of projections can help a GIS user select a projection to best minimize distortions that would degrade a particular type of analysis.

Geocoding

Point geographical features may be positioned by one of two methods. The most common method is through an explicit reference to an (x,y) coordinate location. However, mapped features can also be located implicitly by placing them approximately where they should lie according to their postal address or census tract. This requires that a reference layer exists such as a road network from which the location of the address can be inferred. The process of tying a database of attributes to locations on a map by some addressing method is called *geocoding* and is often used in business and city planning applications.

METADATA

Providing critical information about spatial datasets can reduce misuse and improve access. Therefore, a thorough description called *metadata* (data about data) should accompany each spatial dataset in document, tabular, or database format. Entities have begun to more effectively document spatial datasets by promulgating standards for recording metadata. For example, in the U.S. the Federal Geographic Data Committee (FGDC) has developed a metadata standard that is used by all federal agencies that distribute data to the public (http://www.fgdc.gov/metadata/metadata.html). The required sections in FGDC-compliant metadata are:

1. *identification*: point of contact, data description, date, spatial domain, and access/use constraints - how the data can be used and by whom.
2. *data quality:* both coordinates and attributes, scale to be used, and lineage.
3. *spatial data organization*: description of the data structure.
4. *spatial reference*: ellipsoid, datum, units, projection, and vertical datum.
5. *entities and attributes*: measurement resolution, process of derivation, domain of values, and time period represented.
6. *distribution*: dataset format and access information.
7. *metadata reference*: who prepared the metadata, when, and how.

Although FGDC metadata standards are strict, GIS users should make a habit of completing metadata as fully as possible because the time required to do so

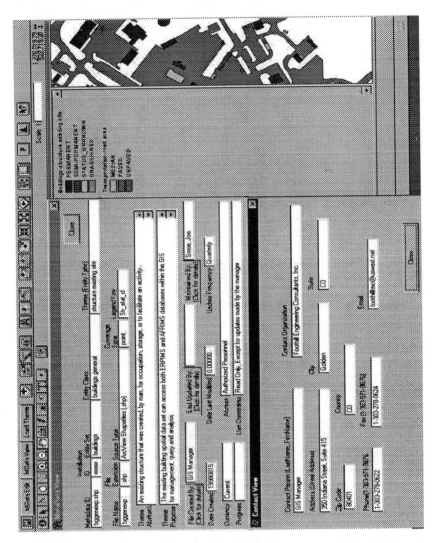

Figure 4   Example of a custom tool for developing and viewing metadata and loading themes using the Tri-Service Spatial Data Standards naming conventions (Courtesy of Foothill Engineering Consultants, Inc.)

will eliminate more time wasted later trying to determine a spatial dataset's georeferences and appropriate use. Metadata ensures that when data are transferred or the originator leaves the organization, each dataset will retain its full value for those who could profit from its use. Poorly documented data can cost an organization, not only in time-wasted, but also in poor decisions made based on GIS results (if incorrectly used in GIS analysis). Tools for developing metadata, if not already included in your particular GIS, can be downloaded at

URL: http://geo-vt.uvm.edu/cfdev2/library/meta/metaweb.cfm and other websites. Figure 4 depicts a tool available as an ESRI ArcView® GIS extension for developing metadata for GIS layers for military installations.

Metadata is invaluable to reduce the cost of finding sources and acquiring spatial datasets. Data dictionaries allow users to search for spatial datasets of interest by keywords listed in the metadata. In the U.S., the National Spatial Data Infrastructure (NSDI) provides access to nearly 100 spatial data clearinghouse nodes on the Internet for various government agencies and private entities to improve public access to spatial data resources around the world. Many state and local governments and even some private entities have adopted a similar metadata-based spatial data access structure on the Internet for their constituencies.

GIS OPERATIONS

It is not possible to describe all of the tools and methods available in GIS for developing, exploring, and analyzing spatial data. However, a brief summary of some of the most common GIS functions can help users grasp the broad utility of GIS. Low-cost GIS are only slightly more useful than automated mapping packages by permitting spatial queries, but little spatial analysis. High-cost GIS offer almost all of the options listed below. For any capabilities not offered in a particular GIS, third party extensions are often available off-the-shelf to extend its functionality. Custom applications can also be developed.

Data Input and Processing

Data Input. GIS provide tools to convert data from different government and commercial formats into the software's own spatial dataset format. Projection, re-sampling, and generalization tools are usually available. GIS also include functions to mosaic and append maps and databases, or clip an area of interest out of a larger map.

Data Validation. Mapped data brought into GIS often includes errors, such as 'leaking' polygons that do not close and lines that overshoot or undershoot a juncture, so tools are included to check and correct a number of topological errors. Methods to simplify editing DBMS and raster cells are also available.

Database Connectivity. Although many GIS include their own DBMS, most GIS today can also directly connect to remote DBMS. This allows GIS to utilize data developed by different departments or at different organizations. Open database connectivity (ODBC) drivers are supported by most GIS software to allow them to connect and read different database formats directly. Dynamic Data Exchange (DDE) is a standard protocol to allow Windows-based applications to exchange data. Standard query language (SQL) support allows GIS to perform information queries on remote databases.

Image Processing. Image processing tools are becoming an important extension of traditional GIS for RS data. Registration and transformation tools are commonly included in GIS to layer images with grid and vector data in a single coordinate system. Visualization and enhancement tools help accentuate, categorize, and extract features from raw images. Error analysis tools can assist in determining how differing quality of each layer can affect the results of multi-image spatial modeling. This can help state limits on accuracy and reliability of results, so conclusions drawn and used in decision-making can be appropriately qualified. Change and pattern detection can be an important use for images in GIS, which can enhance the information available in vector or raster data layers. Fuzzy logic is becoming popular to allow changes to be expressed as a continuum, rather than classified (Jensen et al. 1997).

Data Analysis

Traditional Analysis

Query. GIS excels at answering questions we have about geography through query tools. Although still today most spatial queries are asked in GIS in an almost mathematical way, this should change in the not-so-distant future for several reasons. The object-oriented model discussed in the *Spatial Data Structures* section should allow questions to be asked directly of every-day geographic objects, rather than their dissected representations in relational DBMS. With the rapid developments in voice-recognition software in recent years, rule-based systems are being developed to improve the computer's ability to understand language. Therefore, GIS in the future should be able to accept verbal input and know how to handle a *what*-question differently from a *where*-question automatically.

Topological Tools/Buffers. GIS contain tools for determining distance, proximity, direction, adjacency, and connective relationships between mapped features. Oftentimes, GIS is useful to determine all the objects of one class within a specific distance of another type of object, such as all students within one mile of a new school. GIS is also commonly used to create a *buffer* of a specified distance around an object (Figure 5). An example would be to delineate a vegetative buffer strip of a designated width along the banks of a stream to protect it from degradation in water quality due to unfiltered runoff from adjacent developed lands. The most interior region of a polygon can also be determined for planning hazardous waste storage furthest from neighborhood land uses.

Overlay (Synthesis) of Thematic Layers. In addition to simple queries, GIS excel at answering analytical questions about the complex relationships between numerous thematic layers. This GIS capability is used in site suitability analysis, for extracting dominant trends, and for investigating different solution scenarios to complex spatial problems. Before stacking and analyzing layers as a group, each layer is usually preprocessed separately to convert the raw data into a more

 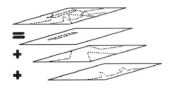

Figure 5 300-ft stream buffer   **Figure 6** Schematic representation
of vector overlay

informative layer by grouping, classifying, rescaling, buffering, and so on. Then the layers can be spatially overlaid to provide information on many different variables at each location. Using map algebra in raster GIS, combinations of mapped variables can be developed in intermediate processes to generate new variables for further model building and data analysis. In vector systems, overlay of polygon layers creates new areas from their intersection (Figure 6). Values in the combined vector layer can then be reclassified to generate homogeneous regions by dissolving boundaries between polygons within a single class. In most cases, multi-layer calculations are summarized into a 'results' map to reveal something new, such as how much better one area would be for development than the rest.

Logical Operators. GIS can develop IF/THEN and AND/OR, and even more complicated logical relationships in analyzing spatial datasets. This is useful for developing complex spatial models that evaluate data differently based on position. For example, the infiltration rate calculations could be improved by developing a spatially distributed model based on soil type and land cover. That is, IF (type x soil, and type y land cover) THEN use (constants x1 and y6) in the equation.

Application-Oriented Analysis

Statistical Analysis. A common use of GIS is to produce summary statistics (mean, median, range, etc.) of a field or fields from the related DBMS based on a selected set of geographical features. GIS is also used to generate regression models from stacked raster datasets to spatially predict a variable based on a number of independent variables included as layers in the stack. Probability tools are used to determine values for missing data in RS images and grids. Probability can also be used to develop fuzzy models and risk assessments in GIS; for example, to determine the placement distance of a new housing development from an endangered species habitat to produce less than a 10-percent chance of causing a decrease in its population. Many GIS include equal-interval, scaled, normalized and quantile map classifications, and histogram charting of spatial datasets to visually investigate how data are statistically distributed. The relationship between two raster datasets can be explored by cross-tabulation and statistical indices that measure dependency and correlation as well. For advanced analysis,

GIS data are exported to statistical packages or the two systems are combined into a single applications framework through off-the-shelf extensions.

Network Analysis. GIS often include algorithms for determining shortest or least-cost path routing, allocation of resources to regions of use, traveling salesman routes to visit several sites most efficiently, and stream ordering. Stream ordering and similar calculations establish contributory and dominance relationships between features in a single spatial dataset. Chapter 5 explains how linear referencing systems can be implemented for a variety of other purposes.

Logistics and Tracking. Large trucking firms use GIS frequently for vehicle routing and scheduling, geocoding addresses, and assigning customer orders to vehicles. GPS on vehicles allows for real-time vehicle tracking in GIS. Tracking tools can also be used to investigate bird migration routes, weather developments, and other phenomenon.

Spatial Modeling and Geostatistics. GIS is popular for generating contour maps or DEMs from interpolation of point data. The results can be overlaid for multi-layered analysis. Splines, inverse-distance weighting, and polynomial surface fitting are popular interpolation methods for specific purposes. One particular method, kriging, serves as the basis for Geostatistics. Kriging attempts to determine autocorrelation between pairs of observed points to uncover a pattern in correlation with distance. The results are used to better model distributed phenomenon. Many variations on kriging have been developed, including co-kriging which uses a second variable related to the first (in GIS, a second thematic layer) to help establish spatial pattern better when data are sparse.

Temporal GIS and Time Series Analysis. As TGIS methods improve to explicitly define temporal aspects of spatial datasets, an impressive array of tools are being developed to analyze land use change over time, the development of natural disasters, and weather phenomenon to predict future events. Population growth effects on natural resources over different time spans may also be investigated with TGIS to establish links and begin to deal with population impacts before they become too severe.

Risk Assessment. GIS has proven useful in risk assessment. For example, GIS has been used to determine risk prices for determining insurance premiums based on location, business type, and hurricane intensity and duration probabilities in hurricane-prone areas (Thomas 1998). GIS can also help determine risk-based corrective action (RBCA) for mitigating hazardous wastes that enter the environment by considering all possible exposure pathways and the potential severity of exposure in a spatial context.

Display/Visualization

GIS include sophisticated mapping, graphing, charting, spatial statistical summaries, and reporting methods to elucidate spatial information resulting from data integration and analysis. Integration of sound, video, photos, and other multimedia presentation methods are also becoming popular in GIS to enhance its information integrating capabilities. Hierarchical data structures or masking methods can allow GIS to display different levels of map generalizations at different map scales. For example, only major highways may show at a 1:100,000 scale, but when zooming into a city at a 1:1000 scale, city streets may become visible. Viewshade (line-of-sight) analysis and shading allow 2.5-D perspectives of mapped features draped over TINs to show relief rising from a two-dimensional plane for viewing from any angle the user chooses. One of the most popular new visualization methods is 3-D rendering combined with animation to allow an observer to view spatial analysis results as if traveling through both time and space in a Disneyland ride. Hazardous waste managers appreciate this capability, using it to analyze the complex geology, groundwater flow, and contaminant characteristics that control contaminant plume development and transport (C Tech Development Corp. 1998). Methods are being developed to represent time as a fourth dimension using texturing and complex perspective techniques (Faust and Star 1997).

INTEGRATION

GIS Object Components in Other Applications

GIS no longer runs only in software specifically designed for spatial data analysis, but through the development of the component object model (COM). Now a particular GIS function can be isolated and embedded in almost any other program as a control code. The most common method (although not the only one) for embedding GIS in windows-based applications is through an OLE/ActiveX custom control (OCX) coded into any compliant development environment such as Visual Basic, Visual C++, or Delphi. For similar cross-platform implementations using the JAVA development environment, the Common Object Request Broker Architecture (COBRA) has been developed. Building GIS into web applications is becoming popular using GIS software providers extensions for this purpose or new languages (learn more about this in the *Societal GIS* section presented later in this chapter).

Integration with Analytical Models

Analytical models that have been integrated with GIS include expert systems (ES) and mathematical optimization. Expert systems take advantage of your computer's capacity for storing data in an organized way. In particular, ES provide a set of rules for the computer to follow to analyze data in much the way an expert would dissect a complex problem. The more rules programmed, the more adept ES is to handle exceptions and unusual circumstances that develop in the problem-solving process. GIS is being integrated with ES for analyzing complex spatial relationships and solving problems based on complicated distributions of phenomena. ES serves as the basis for developing the object-oriented GIS and DBMS systems previously discussed. ES can also be used to represent space at different levels of abstraction (Mukerjee and Hernandez 1995). Artificial Intelligence (AI) may extend the capabilities of ES, if a computer can be programmed to develop new relationships from existing data and rules to mimic human reasoning.

Mathematical optimization is a general term that describes a wealth of computer algorithms designed to solve complex problems in an optimal way. Linear and nonlinear programming, dynamic programming, neural networks, genetic algorithms, and multi-criteria decision analysis (MCDA) can be used to save companies time and money by analyzing criteria and solution scenarios in a formal way. Mathematical optimization techniques attempt to maximize objectivity, so they can be an effective way to incorporate competing requirements to develop a best compromise solution that promotes cooperation. Practitioners have only begun to combine mathematical optimization with GIS, but great opportunities remain for improved decision-making using their combined power.

Spatial Decision-Support Systems (SDSS)

One advantage of GIS is its ability to incorporate advances in technology and methods from a variety of disciplines rapidly. This does not require that all new tools have to be included in a single software package, but merely that multiple software systems are aligned to work together in a single problem-solving environment. The problem-solving environment must allow the user to make decisions without having to be an expert in GIS or any other computer software. In this way, decision-makers and departmental specialists do not have to rely on a computer analyst to assist in solving problems. Effective SDSS are often available in off-the-shelf commercial formats, although custom applications development remains popular. In either case, the software must be error-free and seamlessly integrate GIS and other software with all data resources in a user-friendly system that guides users through the decision-making process while protecting them from making gross input errors or analysis mistakes. Although this was once a tall order, SDSS are becoming commonplace as development and integration tools improve.

Example 1: Automated Mapping/Facilities Management and GIS (AM/FM/GIS). Utilities need sophisticated tools for integrating monitoring systems using automatic supervisory control and data acquisition (SCADA) systems with customer and land base data. AM/FM/GIS often permit real-time mapping of data collected to allow actual results to be compared to modeled predictions for model calibration. Simulations of what would happen in particular emergency scenarios can also be developed (Seifried 1994). The GIS component permits dispatchers to better manage their work crews throughout the city, and develop long-term system upgrade plans rather than operate on crisis management. AM/FM/GIS is also useful in outage management, to analyze the spatial aspect in the pattern of calls received to determine which device or devices probably caused the outage and how this will effect the system downstream (Batty 1998). AM/FM/GIS systems usually contain an object-oriented structure to control each object's behavior explicitly as described previously in the Spatial Data Structures section. More details on AM/FM/GIS are presented in Chapter 8.

Example 2: Land Information Systems (LIS). LIS are specialized GIS for managing data on land and land use. Cadastral systems for organizing and managing real estate often require more accurate parcel boundaries and features than the GIS used for other purposes. For this reason, specific cadastral tools and application packages have been developed for building precise polygon topology from hard-copy records. A DBMS specifically designed for land records information is included along with tools to rationalize the tax base by removing anomalies and inaccurate assessments (Bowden 1997). Active Response GIS (AR/GIS) takes LIS systems one step further by providing a distributed resource modeling environment for non-technical users to interactively develop and evaluate alternative land use scenarios (Faber 1997).

OTHER IMPORTANT ASPECTS

Accuracy, Precision, and Error Propagation

In addition to map distortions produced in the process of map projection, many other types of errors can reduce the utility of GIS. Paper maps are often digitized and then a small area of the digital replica is enlarged beyond the scale of the original. The resulting map can be inappropriately general and details can be inaccurately represented. Digitizing errors may result from technician fatigue or inexperience, closely spaced features, faint linework, etc. Some error is still present even if maps are scanned, as paper shrinks and swells with changes in humidity. Boundaries and transitions between appended map sheets are often incompatible, and in the edge-matching process, features are sometimes moved to incorrect positions. In collecting data to make a map, observations may not be dense enough for the intended scale.

Errors in keying in database attributes and the positions of geographical features can further reduce the accuracy of spatial datasets. Transforming a dataset from one format to another also introduces error. In processing image data, areas obscured by cloud cover are often misclassified. Sometimes, positional errors cause layered spatial datasets to not be related to one another as they should; a building may be shown lying in a lake. During analysis and modeling in GIS, errors are often propagated and compounded in resulting layers. This is especially a problem when spatial datasets of vastly different scales are used together, because the analysis may only be as good as the worst dataset, which may differ significantly from the quality of the best dataset. On the other hand, reducing the error is expensive, and a GIS used to estimate demographic trends might not need to be as accurate as one used to manage cadastre.

GIS tools are being developed to assist users to estimate error in spatial analysis results. *Fuzzy* spatial models are used in some instances to estimate the probability that a feature or a raster cell is correctly represented in both the map and the attributes. Then data that are less certain can be modeled to have less impact on the final analysis results used in reaching important conclusions. Sensitivity analysis and error modeling are also useful methods to determine how errors can impact the spatial analysis results and the decisions based on them.

GIS Implementation

Understanding GIS technology alone is not enough to implement an effective geographic information system. For this reason, the following sections provide insight into the components and methods needed to institute GIS throughout an organization and to better provide GIS services to the clients and the society at large.

An Enterprise-Wide Endeavor

Although expert opinion varies, approximately eighty-five percent of all problems contain some spatial aspect, so data have gained an important position as a corporate asset. Therefore, many organizations are adopting GIS as a way of doing business. GIS today is most often implemented on an enterprise-wide, not a departmental basis, and it usually involves reengineering the way many procedures are performed throughout an organization. Despite the cost and time that instituting a program of this scope usually entails, the adoption of enterprise-wide GIS by organizations has only accelerated because of many important benefits. These benefits can be placed into four main categories:

1. better data integration, accessibility, and enhanced utility
2. reduced costs and added efficiency
3. better integration of operations and collaboration with outside entities
4. better customer service and expanded services.

The spread of the GIS technology occurs in four steps: initiation, contagion, coordination, and integration (Foote and Lynch 1997). Implementing GIS in an organization requires much the same procedure. Everyone in the organization, from top management on down, have to grasp the vision of how GIS will revolutionize their operations.

Initiation. Initiation usually begins with staff in the planning, engineering, or environmental departments, whose necessity to integrate data from all departments in an efficient and cost-effective manner leads them naturally to GIS. However, the department finds that it cannot obtain nor maintain the spatial datasets it needs without collaborating with other departments. Various sharing mechanisms are implemented that may reveal serious problems of duplication, incompatible data sources, and frustration at the amount of time unsuccessfully expended trying to integrate without an enterprise-wide approach.

Contagion. In the contagion phase, an organization learns that enterprise-wide GIS can eliminate many of its most pressing operations and customer service problems. During this period, both management and staff start to become educated in GIS fundamentals and how specifically it has helped other agencies in a similar situation to excel. In fact, the initiation of GIS in surrounding communities is often the impetus for a local government to begin to consider it as a viable alternative for its own operations (Jones 1997). The success of internal pilot projects and experiments with GIS at the project or department level can also produce powerful incentives to move into enterprise-wide GIS.

Coordination. The coordination phase can only begin after management is fully sold on the concept of enterprise-wide GIS and becomes willing to institute the structural changes and provide the financing needed. Often, at this point, an outside GIS consultant and/or GIS software provider are contracted to produce an enterprise-wide GIS implementation plan based on a needs assessment of all departments at all levels. In other cases, a GIS program manager is hired (or designated) to conduct further pilot projects to prove the worth of GIS, and then assist with enterprise-wide adoption and training.

Although some plans are informal, strategic plans are best documented to provide a top-down, long-term vision to achieve enterprise-wide GIS (Somers 1998). The plan must explicitly outline the extent and methods of data and procedural integration between departments. It must establish hardware, software, technical and managerial training, and protocols to achieve success. Security and networking issues must also be adequately address. A time frame must be set and adhered to for each component to be adopted until the system is complete. Staff at all levels must be further educated in how enterprise-wide GIS will change their responsibilities and methods.

A network of support must be established to keep users satisfied with their personal success in the new system. It is usually best that at least one person in each department receives more training and reserves time to serve as a GIS coordinator (Goldstein 1997). An assistant GIS coordinator should also be designated to serve when the GIS coordinator is absent or to take over in case the GIS coordinator should leave the organization. GIS coordinators must train all department staff to maximize efficiency by incorporating GIS into their daily routines. Human resources are expensive to develop, so it is important to target advanced GIS training to the most dedicated employees that are likely to stay with the organization for several years. Since GIS specialists are in high demand, it is likely that some employees may be drawn away soon after receiving training by higher salaries. Therefore, many organizations are requiring employees to sign an agreement that if they leave the organization within a year of receiving specialized GIS or other training, they must repay the training costs. Another approach is to raise employees' salaries that have taken on added responsibilities and/or improved their efficiency after receiving training.

Integration. After a thorough need assessment to determine system requirements and concrete goals, the GIS program can begin to expand, under an adequate financing mechanism, of course. Some of the first steps include evaluating and purchasing hardware and software, setting up a local area network (LAN) if one is not in place, installing access and security mechanisms, hiring expert staff and/or outside consultants (if internal staff are not adequate), and securing a variety of training. The most costly and time-consuming aspect is usually data development, which includes database design and spatial dataset creation. The database design is a particularly critical issue because a poor design limits its usefulness and accessibility. It is often practical to learn from other organizations and their consultants as to what they have done and found productive, and anything they would have done differently if they could. Because of the high cost of making errors in the early stages, prototypes are often generated for general use and feedback before a final system is instituted. Pilot projects are also effective in determining what does and does not work for a particular organization and its culture.

Spatial dataset creation begins by establishing priority layers needed by the greatest number of users. Base layers usually include roads and transportation, parcels, buildings and structures (or at least an aerial photo backdrop), political boundaries, surface water, land use and land cover, zoning, topography, and soils. Other useful layers may include zoning, census tracts, postal codes, parks and landmarks, place names, habitat and species maps, emergency evacuation plans, weather, and increasingly, RS images. Although the federal government has data available in digital format, local governments usually find that detailed requirements dictate that they develop most of their spatial datasets from scratch. Departments have to agree on a consistent measuring system of units, map projections, and scale. In some cases, maps of different resolutions have to be included for the same feature, along with careful metadata descriptions and training in what constitutes appropriate use of a map at each scale.

In all stages of development, and at periodic or need-based intervals thereafter, an enterprise-wide GIS system should be evaluated. Mechanisms are needed to insure that all complaints are duly recorded and investigated, so that the maturing system does not leave any staff members feeling disillusioned or disenfranchised. A testing strategy should be developed in the Implementation Plan for this purpose and carefully exercised. The evaluation should analyze how well the GIS meets accuracy and capability requirements for each user and department, how much improvement in efficiency and effectiveness can be measured at each stage in the development, and how well resources have been reallocated (Chrisman 1997). Understand that GIS may not necessarily decrease agency costs, but free staff and resources to complete more useful work and more community services. Therefore, evaluating enterprise-wide GIS must be conducted at both a cost-effectiveness level and a mission-effectiveness level.

Societal GIS

Today we find ourselves in the midst of the information age, in which information has become an element of the infrastructure required to operate the modern economy (Chrisman 1997). As a tool to integrate data and synthesize information, GIS is itself becoming an important part of everyone's daily lives. For this reason, equitable access to GIS is an important issue. Local government finds this a particular dilemma, because unlike federal government, they usually do not have the resources or the direct mandate to provide public access to data and information free of charge.

In addition to providing spatial information upon request, there are several other reasons why organizations may want to introduce their constituencies to GIS. GIS is an excellent tool for ensuring that decisions are made on a basis of facts rather than for political or emotional reasons. For example, in siting an undesirable land use, GIS can efficiently include every single parcel in the analysis to prove that the final selection is truly the optimal site based on a wide range of criteria and public input. This can often calm opposition, because each party can be certain that its competing concerns have been included in an objective matter. By being able to integrate information from various disciplines, GIS can help resolve conflicts caused by different perspectives, as each group begins to see more clearly how their fields of science interact. This can lead to effective compromise solutions for a necessary public action. When compromise is not the best solution, GIS can often make this clear as well.

Distributed Geographic Information (DGI). The Internet is becoming one of the best ways to distribute GIS to the public, so Distributed Geographic Information (DGI) applications are rapidly gaining popularity. Whether to meet a requirement, allow those who cannot afford the software to enjoy critical GIS services, recover system development costs, or actually make money, DGI may be an excellent compliment to any program (Plewe 1997). Not only can DGI serve outside clients, but also internal clients at scattered offices or out in the field through a wide area network (WAN). DGI can obviously lead to a significant increase in opportunities for capturing, using, and sharing a wealth of geographical data and information.

In developing DGI, an organization must carefully decide what types of data to share, who to share it with, how to distribute the data, how to recover costs or make profit, and how DGI can align with the goals and mission of the organization. DGI services can be divided into seven categories:

1. downloadable raw data archives
2. static map display
3. metadata archives with search tools for locating specific spatial datasets
4. dynamic map browsers that allow for simple data exploration and/or map-making
5. data preprocessors for converting data online
6. web-based query and analysis that incorporates some sophisticated GIS tasks

7. net-savvy GIS software server that provides access to full GIS functionality over the Internet, perhaps on a pay-per-use or pay-per-function basis (Plewe 1997).

An organization could potentially create DGI applications on several levels for different users. For example, types 1 through 3 could be made available for educating the general public. Types 4 and 5 might be made available to other entities with which an organization partners in GIS development. Type 6 might be available on a subscription basis to businesses that profit from the organization's information, while type 7 might be instituted for the organization's technicians who make service calls.

Of course, DGI does require a concerted effort to develop effectively. A multi-disciplinary team is needed and probably should include GIS professionals, information systems staff, a webmaster, web application developers, administrators, sales and marketing staff, public affairs personnel, and legal advisors to handle copyright and use-at-your-own-risk clauses. In the process of developing DGI services, some technical issues will almost inevitably arise. These include storage and transfer of large datasets, hardware/network speed, cross platform problems, security, too many users trying to access the website at once, and users who constantly want website services to be expanded (Calkins 1997).

Many GIS software providers have created tools that simplify the development of Internet Mapping applications. Web languages themselves including the JAVA development environment and standard generalized markup languages (SGML) are also expanding to simplify DGI applications development. For example, the eXtensible Markup (meta) Language (XML) can describe and group data together by allowing users to define markup tags in a Document Type Definition (DTD). Vector data can be incorporated through versions of XML called the Vector Markup Language (VML) or the Precision Graphics Markup Language (PGML) (Gottesman 1998).

Ethics. Ethical issues can arise from expanding your GIS to include DGI. Security is an issue, although encryption, firewalls, and other options can help protect your system. The legal issue of free access may be cause for consternation because an organization may need to recover costs from those who directly profit from their information while also trying to maintain access for those with little ability to pay for data. Some organizations try to handle this by providing a small amount of information free to individuals, but large spatial datasets at a significant cost to businesses. Others try to develop a tiered fee system based on the purpose for which the data will be used. Copyright laws may not cover spatial datasets, which could lead to legal problems when an organization's data are sold by others or misused.

CONCLUSIONS

Despite the relevant issues, spatial data sharing is becoming the key to efficient GIS development and the road to solving problems beyond the local level. However, spatial data can become something of a Trojan Horse if the giver does not take the care to make it as error-free and well documented as possible. The more care we put into data development, documentation, and appropriate analysis, the more value GIS technology will have overall. This will also make us more confident that the information infrastructure we are building will provide us with the knowledge we seek to sustain our world and face the complex challenges the next millennium is certain to provide.

ACKNOWLEDGEMENTS

I would like to thank David Murray (GIS Coordinator, City of Aurora, CO), Cliff Inbau (GIS Coordinator, Brown & Caldwell), and John Labadie (Professor of Civil Engineering, Colorado State University) for their review and suggestions for improving the draft of this chapter.

REFERENCES

Batty, P. (1998). "AM/FM data modeling for utilities." *Technical Paper.* Smallworld, Denver, CO.

Blaha, J. (1998). "VoCarta." URL: http://www.esri.com/parners/gissolutions/ datria/-datria. html.

Bowden, K. (1997). "NovaLis applications used to streamline cadastral mapping in Jamaica." *ArcNews,* ESRI, 19(4), 24.

Calkins, J. (1997). "Implementing a cooperative geospatial infrastructure." *Summer 1997 Educational Workshop.* Environmental Systems Research Institute, Inc., Redlands, CA.

Chrisman, N. (1997). *Exploring geographic information systems.* John Wiley & Sons, New York, N.Y.

C Tech Development Corp. (1998). "Environmental visualization system." URL: http://www.ctech.com/index.htm.

Dana, P. (1998). "Global positioning system overview." URL http://wwwhost.cc.utexas.edu/ftp/pub/-grg/gcraft/notes/gps/gps.html.

Maguire, D. (1998). "Arc/Info Version 8: Object-Component GIS." *ArcNews,* ESRI, 20(4), 1-2,5.

Faber, B. (1997). "Active response GIS: an architecture for interactive resource modeling." *Proc., the GIS'97 Annual Symp. on Geographic Information Systems,* Vancouver, B.C., GIS World, Inc., Fort Collins, CO.

Faust, N. and Star, J. (1997). "Visualization and the integration of remote sensing and geographic information." *Topics in Remote Sensing 5: Integration of*

Geographic Information Systems and Remote Sensing. J. Star, J. Estes, and K. McGuire (eds.), Cambridge University Press, New York, NY.

Foote, K. and Huebner, D. (1996). "Database concepts." URL: http://wwwhost. cc.utexas. edu/ftp/pub/grg/gcraft/notes/datacon/datacon.html.

Foote, K. and Lynch, M. (1997). "Geographic information systems as an integrating technology: context, concepts, and definitions." URL: http://wwwhost.cc.utexas.edu/ftp/pub/-grg/gcraft/notes/intro/intro. html.

Frank, A. (1998). "Different types of 'times' in GIS". *Spatial and Temporal Reasoning in Grographic Information Systems.* M. Egenhofer and R. Golledge (eds.), Oxford University Press, New York, N.Y.

Galetto, R. and Spalla, A. (1996). "Multimedia technology: a new opportunity for GIS improvement." *Data Acquisition and Analysis for Multimedia GIS.* L. Mussio, G. Forlani, and F. Crosilla (eds.), SpringerWien, New York, NY.

Goldstein, H. (1997). "Mapping convergence: GIS joins the enterprise." *Civil Engineering*, ASCE, 49(6), 36-39.

Gore, A. (1998). "The digital Earth - understanding our planet in the 21st century." *Keynote Address at the Grand Opening Gala of the California Science Center in Los Angeles,* February, 1998.

Gottesman, B. (1998). "Why XML matters." URL: http://www.8.zdnet.com/ pcmag/-features/xml98/index.html.

Jensen, J., Cowen, D., Narumalani, S., and Halls, J. (1997). "Principles of change detection using digital remote sensor data." *Topics in Remote Sensing 5: Integration of Geographic Information and Remote Sensing.* J. Star, J. Estes, and K. McGwire (eds.), Cambridge Univ. Press, New York, N.Y.

Jones, T. (1997). "Nassau County: model for enterprise GIS." *ArcNews,* ESRI, 19(4), 10-11.

Krzanowski, Paylylyk, and Crown. (1993). "Glossary of GIS terms." *1994 International GIS Sourcebook*, GIS World, Fort Collins, CO.

Mukerjee, A. and Hernandez, D. (1995). "Tutorial on representation of spatial knowledge." *Fourteenth International Joint Conference on Artificial Intelligence,* Sunday, Aug. 20, 1995, Montreal, Canada.

Plewe, B. (1997). *GIS on line - information retrieval, mapping, and the Internet.* OnWorld Press, Santa Fe, NM.

Seifried, C. (1994). "Merging GIS and SCADA." *Amer. City & Co.*, 8(6), 42-47.

Somers, R. (1998). "Ensuring full adoption of GIS." *Geo Info Systems*, (May 1, 1998), 46.

Thomas, R. (1998). "Reinsurance researchers model risk with ArcView GIS." *ArcUser*, ESRI, 2(1), 29-30.

Trimble. (1995). "Aspen Pro - sub-meter, map-based data capture for GIS." *Surveying and Mapping Products Brochure*, Trimble Surveying and Mapping Division, Sunnyvale, CA.

Trimble. (1996). "GeoExplorer II - pocket-sized GPS mapping system." *Surveying and Mapping Products Brochure*, Trimble Surveying and Mapping Division, Sunnyvale, CA.

CHAPTER 3

Remote Sensing

K. Jeyapalan

This chapter presents material that will help those using geographic information system (GIS) understand the basics of remote sensing. It outlines the history of remote sensing and describes the remote sensing process, including electromagnetic spectrum, electromagnetic radiation, atmospheric windows, and remote sensing platform. Remote sensing devices, ranging from aerial camera and radar beam videocon to microwave sensing, are described. Various approaches to analysis of remote sensing data are discussed, including energy interactions within Earth surface features, spectral response patterns, theory of photo interpretation, analysis of digital Landsat data, and photogrammetry. In the final section of the chapter, the applications of remote sensing in GIS are illustrated by integrating the two.

INTRODUCTION

Remote sensing has been developed from photo-interpretation, which means identifying objects in a photograph. The term was first used by Evelyn L. Pruitt of the Office of Naval Research to mean the observation from remote platforms in the regions of the electromagnetic spectrum beyond the range of human vision and photographic sensitivity (ASPRS 1975). Remote sensing is done from acquisition platforms such as satellites using multi-spectral scanners (MSS) that are called imaging devices. Digital image analysis is a measurement method used to analyze these data.

History of Remote Sensing

The first photograph was taken by the Frenchmen, Daguerre and Niepce in 1839, followed by Petit Bicetree who, in 1858, took photographs of an Italian village (ASPRS 1975). Then in 1906, aerial photographs of the earthquake-destroyed city of San Francisco were taken by a 1000 lb camera hoisted in the air by a balloon kite. Soon after Wilbur Wright piloted the first airplane taking motion pictures over Centocelli, Italy, aerial photography (because of their use in military applications) became very popular in both world wars. Civilian applications followed especially in the area of resource management such as farming and timber. Aerial photography also became popular in topographical and geological mapping.

In 1931, infrared (IR) sensitive film was developed and between 1935 and 1938, Britain developed the IR systems, imaging systems, and detectors to detect aircraft during World War II. The development which contributed to the advancement of aerial photographic technology was invented in 1889 by Heinrich Hertz of the Radio Detection and Ranging System, popularly known as RADAR which was initially used to detect ships at sea (ASPRS 1975). In 1951, Carl Wiley discovered Synthetic Aperture Radar (SAR), the laser was used in 1956 by Russian astronomer N.A. Kozyrev to scan the surface of the moon. The gyroscopically stabilized camera invented in 1907 by Alfred Mauil was used in 1946 to take pictures from V2 rockets. In 1963, Merifield pioneered the geological interpretation from hyper altitude photography. In 1964, Morrison and Chown constructed a series of geological maps from the photographs taken from Mercury spacecraft which orbited the Earth in 1961. NASA developed the first Earth Resources Technology Satellite (ERTS) in 1972. It used a 3-videocom camera and a 4-band MSS as its imaging system. The availability of MSS imaging in digital form and the advent of computer led to the development of image analysis. It is now possible to identify objects from digital image using computer software.

The pixel size, the smallest element of the digital image, of the MSS ERTS-I is about 79m at ground scale. The ERTS–I was called Landsat-I. It was decommissioned in 1978 and replaced by Landsats 2, 3, 4, 5, 6, and 7, the last of which is scheduled to be launched in 1999 with 30m pixel size MSS and 15m ground resolution videocon camera. A French satellite named SPOT was launched in 1986 for Earth resource applications. SPOT has 10m ground resolution in panchromatic imagery and 20m pixel size MSS. Subsequently SPOT 2, 3, and 4 were launched, the last in 1997.

Satellite and aerial imagery in digital form and availabile image analysis software to process them, enable geographic information system (GIS) users to identify changes in GIS base maps and update them. Remote sensing technology does not, however, allow users to locate objects in maps. The science of detecting the location of an object in an image is known as photogrammetry. As in most of the sciences photogrammetry too has developed considerably with advances in computer technology. Prior to the use of soft photogrammetry, photogrammetry

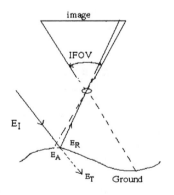

Figure 1   A remote sensing sensor

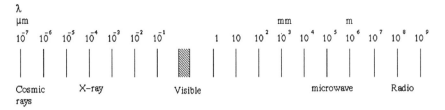

Figure 2   Electromagnetic spectrum

used hard copy film and analogue instruments to locate objects. Using digital imaging, computer software, and a personal computer, the science of soft photogrammetry locates objects on maps and updates them. For more details on the history of remote sensing, the reader is referred to ASPRS (1975).

REMOTE SENSING PROCESS

Electromagnetic Spectrum

Remote sensing equipment consists of a sensor mounted on a platform. It collects energy emitted, reflected, and transmitted by an object and records it on an image plane (Figure 1). The recorded value, described as the response, is the characteristic of the object and it can be used to identify it provided a response signature is available. The energy is then carried in the electromagnetic spectrum which consists of waves with different wavelength, λ, and frequency, f. The velocity of the electromagnetic wave or light, C, is then given by:

$$C = f\lambda \qquad (1)$$

Figure 2 shows the electromagnetic spectrum ranging from $-\infty$ to $+\infty$ in wavelength. The human eye can only detect energy in the visible spectrum. Figure 3 shows different sensors and their spectrum range.

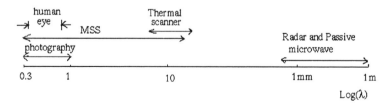

Figure 3   Different sensors and their range in the electromagnetic spectrum

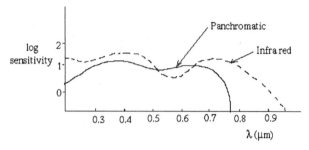

Figure 4   Spectral sensitivity curves

Figure 4 shows the spectral sensitivity curves of a panchromatic (black and white) and an infrared film. The log sensitivity is a function of the response of the detector to the energy in the specific wavelength. The curves indicate that panchromatic films cannot be used to detect objects emitting energy in the infrared regions of the spectrum.

Electromagnetic Radiation

All matters at temperature, T, above absolute zero (°K) continuously emit electromagnetic radiation. The energy radiated is a function of the surface temperature. The energy radiated, M, by a black body is given by Stefan-Boltzman law:

$$M = \sigma T^4 \qquad\qquad (2)$$

where σ = Stefan Boltzman constant. Figure 5 shows a sketch of the spectral distribution of the emitted energy, M_λ, by the sun and the Earth at different temperatures. Figure 5 suggests that the energy from the sun can be detected at about 0.5μm wavelength whereas the energy from the Earth can be detected at about 9μm.

The reflected energy from the sun is used by remote sensing to detect objects. At night when there is no reflected energy from the sun remote sensing uses the energy emitted by the Earth to detect objects. In radar, however, the energy is

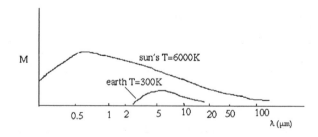

Figure 5 Spectral distribution of emitted energy

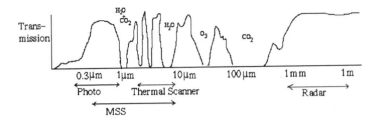

Figure 6 Atmospheric windows

supplied by the sensor and when it does, the sensor is an active sensor. When the energy is supplied by the sun or another source the sensor is in a passive mode.

Atmospheric Windows

Energy travels through the atmosphere which contains water (H_2O), carbon dioxide (CO_2), and ozone (O_3). These gas particles absorb energy within certain ranges of the spectrum. Figure 6 shows transmission spectra of the atmosphere. It indicates the different spectral windows for remote sensing applications. It also shows the types of sensors that are applicable for use in each window.

Remote Sensing Platform

The degree of response of the sensor depends on the intensity of the energy received which, in turn, depends on the distance of the sensor from the object. Figure 7 shows the types of platform used in remote sensing. To uniquely identify objects and changes, remote sensing uses multi-platform data, multi-sensor data, multi-date data, multi-station, multi-direction, and multi-vision data.

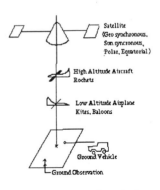

Figure 7   Remote sensing platforms

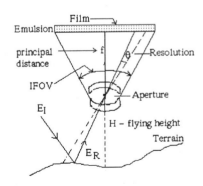

Figure 8   Aerial camera

SENSING DEVICES

Two types of sensing devices are used in remote sensing: thermal detectors and quantum detectors. Thermal detectors, such as bolometer or thermal couple, absorb the radiant flux or the amount of energy per time interval incident on the surface and responds with an electronic signal proportional to the radiant flux. The advantage of the thermal detector is the extended range of its spectral sensitivity. Thermal detectors, however, are not widely used in remote sensing.

Quantum detectors are of two types: photographic and electronic. There are two groups of electronic detectors: photo-emissive detectors such as return beam videocon (RBV) where the photons are emitted into vacuum or gas and solid state detectors such as multi-spectral scanner (MSS) where the exited charge is transmitted within the solid by holes or electrons.

Aerial Camera

An aerial camera consists of an objective lens (Figure 8) that is free of both geometric distortions and aberrations. The aberrations degrade the sharpness quality of the image, whereas the distortions deteriorate its geometric quality. The angular separation, θ, between the objects that can be resolved by the lens is:

$$\theta = 1.22 \, \lambda/A \tag{3}$$

where λ = wavelength of the incident energy and A = aperture of the lens.

The camera also consists of a film, that is coated with silver halide emulsions. The photons from the sun are reflected by the object and focussed on the film by the lens. The emulsion particles, which are about 25-40μm, react precipitating dark Ag particles with density proportional to the radiant flux. Depending on the sensitivity of the film, the picture when developed reveal in black and white (panchromatic) or red, green, and blue (color) images. The quality of the film is measured by the number of lines/mm (lpm) of the object that can be resolved in the imagery. The LPM of a camera is determined by photographing an object consisting of a number of lines/mm and resolving them. The LPM can deteriorate if the camera is mounted on a moving platform such as an airplane. In this event the film must be moved proportionately to compensate for movement caused by the movement of the platform. LPM of the aerial camera is about 50-60.

The scale of the developed image, S, which is the ratio of the image distance between two objects to the ground distance between them, is given by

$$S = f/H \tag{4}$$

where f = principal distance from objective lens plane to film plane and H = flying height (distance from objective lens plane to ground plane) (Figure 8). When the scale is uniform f and H are the same for all objects and the film plane is parallel to the ground plane. In practice, the film plane is made parallel to the lens plane. The ground is not a plane and the film plane is not always perpendicular to the vertical. Therefore the scale of the photograph may not be uniform.

In order to limit the geometric distortions and improve the quality of the images, the size of the film in the camera is restricted by the instantaneous field of view (IFOV) which is the total angular exposure of the film (Figure 8). The amount of energy, E, incident at a point in the film is:

$$E = \frac{sd^2t}{4f^2} \qquad (5)$$

where s = brightness of the scene or the object, d = diameter of the shutter diaphragm, and t = exposure time or shutter speed,. The aperture setting or F/stop or F-number is f/d and the shutter speed should be adjusted to avoid over and under-exposure.

There are three types of films: panchromatic, orthochromatic, and infrared. The panchromatic is sensitive to 0.4-0.8μm (red, green, and blue), the orthochromatic is sensitive to 0.4-0.6μm (blue and green), and the infrared is sensitive to 0.3-0.9μm (all visible light and near infrared). The film can be either black and white or color. There is only one layer of emulsion in black and white film, whereas there are three layers in the color film. Color photography is based on the principle of subtractive color mixture using a yellow dye layer sensitive to blue, a magenta layer sensitive to green, and a cyan layer sensitive to red light.

Return Beam Videocon (RBV)

The RBV does not contain film, instead in it photons are concentrated by the lens on a photosensitive surface. The secondary electrons proportional to the incident photons are stored by the target and a fine wire-mesh screen in front of it. Each point on the mesh is raster scanned every thirtieth of a second by a beam of electrons from an electron gun. The reflected beam containing the secondary electrons or the signal information is amplified by the electron multiplier. The electrons or signals can be used to produce static or dynamic display on the phosphor-coated transparent free plate or cathode ray tube. The signal corresponding to each pixel of the target can be used for image analysis in remote sensing (Figure 9).

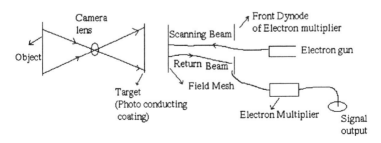

Figure 9   Working of an RBV

39

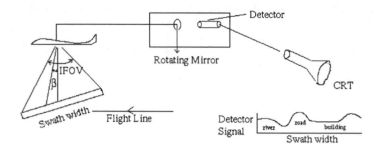

Figure 10 Working of MSS

Because RBVs image an active scene instantaneously in camera fashion, these images can be used for photogrammetric applications. By having a reseau grid on the target plane, an array of tick marks can be imaged. Knowing the image position versus the calibrated positions of these marks, any geometric distortion resulting from recording or transmission can be corrected.

Landsat uses the RBV with a 50×50mm target surface and resolution of about 100 LPM The RBV has an advantage over the photogrammetric sensor. The output signal from the RBV can be transmitted like radio waves or as in the television whereas in the photographic sensor the film must be recovered which is impractical in satellite remote sensing.

Multi-Spectral Scanner (MSS)

The optical mechanical scanner consists of a coating of infrared sensitive material such as copper doped or gold doped germanium on the end of an electric wire. The infrared photons striking the detector generate an electric signal that varies in intensity according to the amount of thermal energy from the terrain viewed by the mirror. The detector signal will vary across the flight line and the swathe width according to the terrain feature (Figure 10). The scanner using properly sensitized materials can provide multi-band imagery hence known as the multi-spectral scanner (MSS).

The MSS used in Landsat has a rotating mirror that moves the field of view of the scanner along a scan line perpendicular to the direction of flight. The forward motion of the aircraft advances the strip that is being viewed between the scans causing a two-dimensional (2-D) image dataset to be recorded.

In the MSS the incoming energy is separated into spectral components that are sensed independently. To separate the reflected wavelength from the emitted wavelength in the incoming radiation, a dichroic grating is used. The reflected wavelength component is directed from the grating through a prism that splits energy into a continuum of UV, visible and reflected IR wavelengths. The dichroic grating dispenses the components of the incoming signal into constituent

40

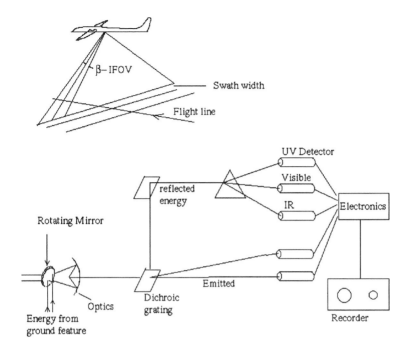

Figure 11 Separation of spectral bands

wavelengths. By placing an array of detectors at the right geometric positions, the incoming beam is split into multiple bands or channels and each measured independently.

The signals generated by each detector of the MSS are amplified by the electronics of the system and recorded on magnetic tape with multi channels. On board the aircraft these data are recorded on high-density tape, and on the ground the tape is transformed on to computer compatible tape (CCT) to allow processing.

Since MSS systems sense energy over a small IFOV (2.5 radians) to optimize resolution, a limited amount of energy is incident on each of the system's detectors. Hence detectors should be very sensitive to output signals that are stronger than the noise (generated by the system such as extraneous unwanted noise). Noise masks signal fluctuations that are weak. The strength of the MSS signal is directly related to the characteristics of the terrain. Large changes correspond to changes in cover type and subtle changes correspond to subclasses in the cover type (Figure 12). The critical factor is the ratio of signal strength to noise level (S/N):

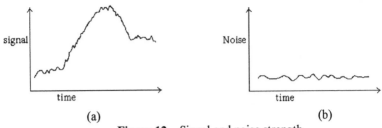

Figure 12 Signal and noise strength

$$\left(\frac{S}{N}\right)_{\lambda} \propto D_{\lambda} \beta^{2} \left(\frac{H}{V}\right)^{2} \Delta\lambda \, L_{\lambda} \qquad (6)$$

where

$D =$ detectivity (measure of detector performance quality)
$\beta =$ IFOV
$H/V =$ flying height/aircraft velocity
$\Delta\lambda =$ spectral resolution
$L_{\lambda} =$ spectral radiance of the ground feature

The output from the detectors can be obtained either as a hard copy or numerical data. The numerical data can be pre-processed and corrected prior to the analysis. Two such corrections are geometric and radiometric corrections.

Microwave Sensing

Microwave energy has the following desirable characteristics:

- depending on wavelengths involved, microwave energy can 'see through' haze, light, rain, snow, clouds, smoke etc.
- microwave reflections or emissions from Earth materials bear no relationship to counterparts in visible or thermal portions of the spectrum.

Radar is an active microwave sensor. Microwave radiometer is a passive sensor responding to low levels of microwave energy that are naturally emitted and/or reflected from the terrain features.

Radar Development

The word 'Radar' is an acronym for radio detection and ranging. It was developed to detect the presence of objects and their range. Radar transmits short pulses of microwave energy and records the strength of signals received from objects. The Doppler radar system uses the frequency shift in the transmitted and reflected signals to measure the velocity of an object.

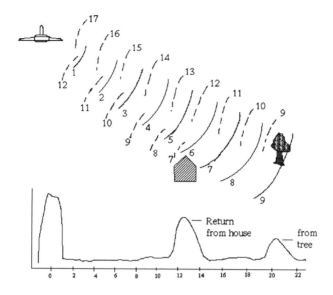

Figure 13   Side looking airborne radar (SLAR)

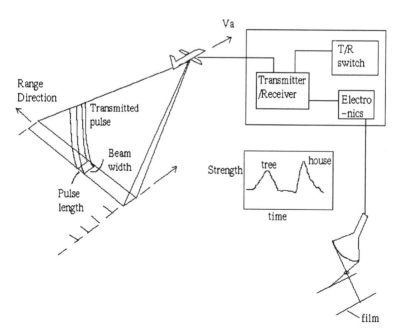

Figure 14   Working of SLAR

The plan position indicator (PPI) system has a circular screen on which a radial sweep shows the position of the radar 'echoes.' PPI radar images a continuously updated plan view map of objects surrounding its rotating antenna. The PPI system is used in weather forecasting, air traffic control, and navigation. Because of its poor spectral resolution, it is not used in remote sensing.

The spatial resolution is determined by the size of the antenna of the PPI system. Generally, remote sensing by airborne radar is done using a system that has an antenna fixed under the aircraft and pointed to the side and therefore known as side looking airborne radar (SLAR).

SLAR Operation Systems

By measuring the return time of the signal electronically, t, the distance to the object or its range, SR, is determined:

$$SR = Ct/2 \qquad (7)$$

As the aircraft advances, the antenna is continuously repositioned in the flight direction and switched from transmitter to receiver mode by synchronizer switch. Transmitted pulse returns echoes from terrain features occurring along the beam width of the antenna (Figure 13). The echoes are received and processed by the airborne antenna to produce video signals (amplitude/time). The signal modulates the intensity of a single line cathode ray tube exposing an image line on the film. Such lines in the image are tonal representatives of the strength of the signal received from a single radar pulse. Between lines the film is advanced at a velocity V_f proportionate to the velocity of the aircraft, V_a. The combined response of these pulses yield a 2-D image (Figure 14).

Spatial Resolution. Ground resolution is controlled by the length of the pulse and the beam width of the antenna. The length of the pulse is determined by the amount of time it takes for the antenna to emit its pulse or energy and it dictates the resolution in the direction of energy propagation. The width of the beam of the antenna determines the resolution in the direction of flight or azimuth.

Range Resolution. To image two ground features that are close to each other separately in the direction of the range, it is necessary for the signal reflected from two objects to be received separately by the antenna. Any time the overlap between signals from the two objects will cause the image to be blurred.

When the slant range distance between buildings is less than half the pulse, then it has had time to travel to B and have its echo return to A while the end of the pulse at A continues to be reflected (Figure 15). Taking into account the depression angle effect (Figure 16), the ground resolution in the range direction, R_R, is given by

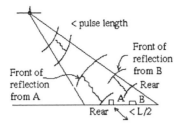

Figure 15   Range resolution

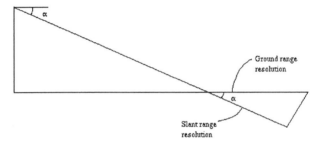

Figure 16   Ground resolution

$$R_R = \frac{C\,\tau}{2\,Cos\alpha} \qquad (8)$$

where τ is pulse duration .

Azimuth Resolution. The azimuth resolution, R_a, is determined by the width of the angular beam of the antenna, β, and the ground range, GR. With increasing distance from the aircraft the azimuth resolution deteriorates. Therefore, objects A and B are resolved at GR_1 and not at GR_2 (Figure 17):

$$R_a = GR \cdot \beta \qquad (9)$$

The beam width is given by $\beta = \lambda/L$, where λ = wavelength of the transmitted pulse and L = length of antenna. β can be controlled by
1. physical length of antenna
2. synthesizing an effective length of the antenna synthetic aperture.

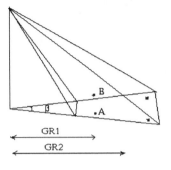
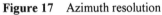

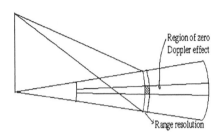

Figure 17   Azimuth resolution   **Figure 18**   Doppler shift

The synthetic aperture employs a short physical antenna. Through modified data recording and processing technique, it can synthesize the effect of a long antenna as well. For example, a 2-m antenna can be converted to make it effectively a 600-m antenna. In essence, return signals from the center of the beam are distinguished from each other by detecting their Doppler shifts (Figure 18). Return signals from features in the area ahead of the aircraft have a higher frequency than those within the wide antenna beam because of the Doppler effect. Conversely, return signals from the area behind the aircraft have lower frequencies. Returns from the center will show no frequency shifts. By processing return signals according to their Doppler shifts, a small and effective beam width is generated.

Both amplitude and frequency of signals from objects within the beam of the moving antenna should be recorded. Because the signals received by a synthetic aperture system are recorded over a long period, the aircraft translates those received from the real antenna over the corresponding distance. The azimuth resolution is essentially independent of range and it is described as being *unfocussed* when the resolution is a function of wavelength and range and not of the antenna length. It is described as being *focussed* when the resolution is a function of the antenna and is independent of λ and its range. For example, a 1-m antenna would give a 0.5m resolution irrespective of the range of the feature.

APPROACHES TO DATA ANALYSIS

Identifying features and their location using spectral response from the sensor depends on:

- the spectral sensitivity of the available sensors
- the presence or absence of atmospheric windows in the spectral ranges within which the sensor operates
- the source, magnitude, and spectral composition of available energy
- the geometric qualities of the sensor
- knowledge about the spectral response patterns of the features.

Energy Interactions within Earth Surface Features

The incident energy, E_I, from the source will be reflected, E_R, absorbed, E_A, and transmitted, E_T, by the feature (Figure 19). Thus,

$$E_I(\lambda) = E_R(\lambda) + E_A(\lambda) + E_T(\lambda) \qquad (10)$$

It should be noted, however, that: (1) the proportions of E_R, E_A, and E_T will vary for different Earth features and (2) E_R, E_A, and E_T depend on wavelength. Even within a given feature type, reflected, absorbed, and transmitted energy will vary at different wavelengths.

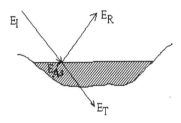

Figure 19 Energy interaction with Earth features

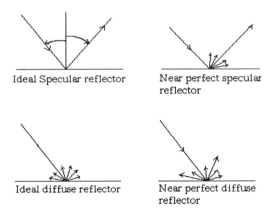

Figure 20 Types of reflectors

Since sensors detect reflected energy, reflection properties of the features are important in remote sensing. Then,

$$E_R(\lambda) = E_I(\lambda) - [E_A(\lambda) + E_T(\lambda)] \qquad (11)$$

The manner by which an object reflects energy can be either specular or diffused. Specular reflectors have flat surfaces that manifest mirror-like reflections. Diffuse or Lambertian reflectors have rough surfaces that reflect energy uniformly in every direction (Figure 20). The characteristics that categorize any given surface is dictated by its condition – whether it is rough or smooth in comparison to the wavelength of the energy incident upon it. For example, in the long wavelength portion of energy, rock terrain will appear as smooth, whereas in the short wavelength portion, it will appear rough.

Diffused reflectors contain spectral information about the surface whereas specular reflectors do not. Thus in remote sensing, we are interested in diffuse reflectance properties of features. The spectral reflectance of a feature is defined as a spectral reflectance curve that is a plot of R_λ Vs. λ,

$$R_\lambda = \frac{E_R(\lambda)}{E_I(\lambda)} \qquad (12)$$

In a black and white infrared film, a maple tree (with higher infrared reflectance) will appear lighter than a pine tree. Figure 21 shows the spectral reflectance curves of maple and pine trees. In the range 0.7-1.3μm, a leaf reflects 50% of the incident energy. Within this region, a plant's reflectance results from its internal structure that is variable between plant species. It can be used to identify different species of maple and pine. The factors that influence soil reflectance are:

- moisture (presence of moisture decreases soil reflectance)
- texture (proportion of sand, silt, and clay)
- roughness of surface (iron oxide and organic matter).

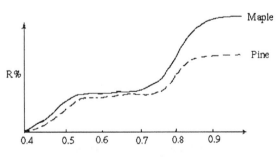

Figure 21   Spectral reflectance curve

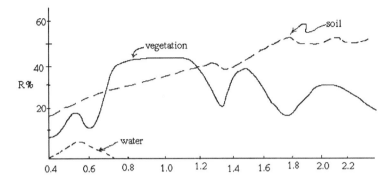

Figure 22   Feature identification by spectral reflectance patterns

In the spectral reflectance of water, the most important characteristic is the energy absorption at reflected infrared wavelengths. Because of this absorption water bodies can be located easily by reflected infrared wavelengths. Figure 22 shows that the water can be identified in the infrared region, vegetation in the 1.0μm region, and soil in the 2.0μm region.

Spectral Response Patterns

In order to identify features, a spectral signature or value has to be developed. Spectral signatures for different features are given by:

- spectral reflectance
- spectral emitance
- radiation measurements.

These characteristics represent response patterns rather than signatures. A signature of a feature implies that it is a unique and absolute value for the feature, whereas a response pattern implies that it is quantitative but not absolute. Spectral response patterns are affected by: (a) temporal effect, (b) spectral effect, and (c) atmospheric effect. In practice, spectral response patterns are developed for each feature.

Theory of Photo Interpretation

The incident energy, $E_I(\lambda)$, is reflected, transmitted, and absorbed by the feature according to (10). The reflected energy, $E_R(\lambda)$, corresponding to sensitive wavelengths creates an image with certain texture (rough/smooth), tone (light/dark), and resolution (Figure 23). The radiometric values recorded in each pixel of the image correspond to $R_R(\lambda)$ and the resolution θ is given by

$$\theta = \lambda/A = d/D \qquad (13)$$

where A = aperture diameter, d = ground size of the object, and D = distance of the object from the sensor.

The photo interpreter, who is familiar with the terrain, develops a key (signature) to identify rocks, grass, soil, trees, man-made features, urban and natural developments and geological features using the texture and tone of imagery. This key is usually field checked and then used to classify imagery. Since this procedure is slow photo interpretation is now done by computers. The image is described by a numerical matrix of brightness (radiometric) values corresponding to $E_R(\lambda)$. These values are obtained either by scanners, such as densitometers or directly from the sensor and quantitatively analyzed by computers (Figure 24).

Analysis of Digital Landsat MSS Data

Since Landsat data are in digital form, they are suitable for computer-assisted analysis. Therefore, a multitude of procedures has evolved to process MSS data on compute compatible tapes (CCT) format. Computer-based procedures for analyzing Landsat data can be grouped into three categories:

- *Image Restoration*: Image restoration is the procedure for correcting a variety of radiometric and geometric distortions that may be present in the original image data.
- *Image Enhancement:* These are the techniques, which are applied to accentuate the contrast between features in the imagery.
- *Image Classification*: These are the quantitative techniques, which are applied to automatically interpret image data. Each pixel information is evaluated and converted to an information category that replaces the image data file with a matrix of category types.

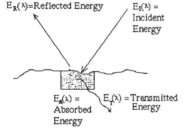

Figure 23 Energy interaction with Earth feature **Figure 24** Matrix of numerical

Radiometric Corrections

Since detector output changes gradually over time, it is necessary to calibrate the data they produce. The detectors are calibrated by (a) viewing an electrically illuminated step-wedge filter during each mirror sweep and (b) viewing the sun during each orbit to provide absolute calibration. These calibrated values are used to develop radiometric correction functions for each detector. The correction functions yield digital numbers that correspond linearly with radiance and are applied to all data prior to dissemination. The radiant values contain an added component due to 'air light' reflected from the atmosphere. The removal of this component is described as 'haze removal.' Occasional problems due to transmission cause a 'striping' effect, which is corrected by a normalizing procedure.

Geometric Corrections

These corrections are needed due to variations in altitude, attitude, and velocity of the aircraft and eastward rotation of the Earth. The eastward rotation of the Earth causes the mirror sweep to view an area slightly to the west of the previous sweep. Geometric corrections, which are random and complex distortions, are made by analyzing ground control points and developing the following functions to transform image coordinates (x, y) to ground coordinates (X, Y):

$$X = f_1(x, y) \qquad\qquad 14(a)$$

$$Y = f_2(x, y) \qquad\qquad 14(b)$$

The process by which geometric transformations are applied to the original data is called *resampling*. Using f_1 and f_2, the appropriate pixel value (x, y) is transferred from the image dataset to the geometrically correct matrix (Figure 25).

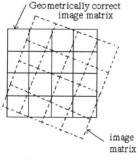

Figure 25 Geometric correction

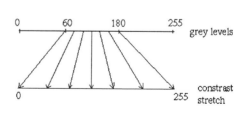

Figure 26 Contrast stretch

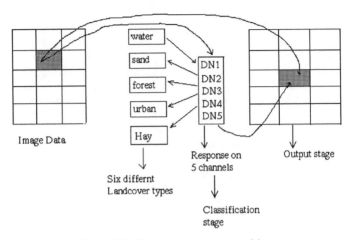

Figure 27 Spectral pattern recognition process

Digital Enhancement Techniques

Most enhancement techniques can be characterized as either point or local operations. Point operations modify the radiometric values of each pixel in an image dataset independently. A point operation enhancement commonly applied to Landsat data is the *contrast stretch* (Figure 26). Local operations modify the radiometric values of each pixel by considering the pixel values that surround it. They are used to either emphasize or de-emphasize abrupt changes in pixel brightness values.

Classification Stage

Quantitative techniques can be applied to automatically interpret digital image data. In this process each pixel observation is evaluated and assigned to an information category, thus replacing the image data file with a matrix of category types. *Spectral pattern recognition* is one of many forms of automated pattern recognition. In spectral pattern recognition, the process illustrated in Figure 27 is termed a supervised process and consists of three stages:

- *training stage* where the analyst compiles an 'interpretation key' developing numerically the spectral attributes for each feature type of interest
- *classification stage* where each pixel in the image dataset is compared to each category in the numerical interpretation key
- *output stage* where each pixel is then labeled either with the name of the category it resembles or as unknown.

Figure 28 Minimum distance   **Figure 29** Parallelpiped classifier

The classification described above is done by the following methods:

(a) minimum distance classifier in which the unidentified pixel is classified into the category to which its radiometric value is closest. Figure 28 shows that the plot of radiometric values of Band 3 and 4 sensors. The unclassified pixel p can be classified as Urban (U) instead of Corn (C) field as p is closer to the centroid of U.

(b) *parallelepiped classifier* in which the pixel is classified to that category if its radiometric value falls within the specified parallelepiped of the category. In Figure 29, since p is outside either of the parallelpiped it can be classified as unknown.

(c) Gaussian maximum likelihood classifier in which the probability f(x) of a radiometric value X in band 3 belongs to category with mean μ and standard deviation σ is given by

$$f(X) = \frac{1}{\sigma\sqrt{2\pi}} e^{-\frac{1}{2}\left(\frac{X-\mu}{\sigma}\right)^2} \qquad (15)$$

The probability f(x) for a radiometric value X in band 4 and Y in band 3 belongs to category of means and standard deviations (μ_x, σ_x) in band 3 and (μ_Y, σ_Y) in band 4 is given by

$$f(X,Y) = \frac{1}{2\pi\sqrt{\sigma_X^2 + \sigma_Y^2 - \sigma_{XY}}} e^{-\frac{\sigma_X^2\sigma_Y^2}{2\left(\sigma_X^2+\sigma_Y^2-\sigma_{XY}\right)}\left\{\left(\frac{X-\mu_X}{\sigma_X}\right)^2 - 2\sigma_{XY}\frac{(X-\mu_X)(Y-\mu_Y)}{\sigma_X^2\sigma_Y^2} + \left(\frac{Y-\mu_Y}{\sigma_Y}\right)^2\right\}} \qquad (16)$$

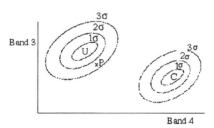

Figure 30 Gaussian maximum likelihood classifier

The major, a, minor, b, and the orientation, γ, of the probability ellipses are given by

$$a^2 = \frac{1}{2}\left(\sigma_X^{\,2} + \sigma_Y^{\,2}\right) + \sqrt{\frac{1}{4}\left(\sigma_X^{\,2} - \sigma_Y^{\,2}\right)^2 + \sigma_{XY}^{\,2}} \qquad (17)$$

$$b^2 = \frac{1}{2}\left(\sigma_X^{\,2} + \sigma_Y^{\,2}\right) - \sqrt{\frac{1}{4}\left(\sigma_X^{\,2} - \sigma_Y^{\,2}\right)^2 + \sigma_{XY}^{\,2}} \qquad (18)$$

$$\tan 2\gamma = \frac{2\sigma_{XY}}{\sigma_X^{\,2} - \sigma_Y^{\,2}} \qquad (19)$$

Figure 30 shows the probability ellipses of U and C for Band 4 and 3. The pixel, p, can be classified with U with probability between 90-95% confidence level. This process can be expanded to the classification of multi-band data using:

$$f(X_1, X_2, \ldots X_n) = \frac{1}{(2\pi)^{\frac{n}{2}}\sqrt{|\Sigma|}} e^{-\frac{1}{2}(X - \mu_X)^T \Sigma^{-1}(X - \mu_X)} \qquad (20)$$

where $(X - \mu_X)$ is the vector of radiometric values and Σ is the matrix of covariance of the radiometric values.

Output Stage

After all the data have been categorized, the results are presented in the output stage, commonly in the form of a map, which can be used in GIS. There exists a

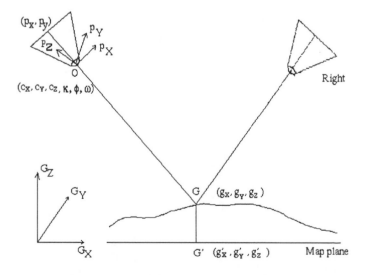

Figure 31   Coordinate systems in photogrammetry

number of software, such as Imagine (Erdas) that can classify the dataset and produce the map.

Photogrammetry

The science of locating the positions of a feature from the imagery is known as photogrammetry. Geometrically speaking the photograph or the imagery is a central projection of the ground features where the center of the objective lens is the center of the projector. The coordinates (p_x, p_y) of an image I in the image coordinate system (p_X, p_Y, p_Z) (see Figure 31) is given by the central projection equation

$$p_x = \frac{a_{11}(g_X - c_X) + a_{12}(g_Y - c_Y) + a_{13}(g_Z - c_Z)}{a_{31}(g_X - c_X) + a_{32}(g_Y - c_Y) + a_{33}(g_Z - c_Z)} \qquad (21a)$$

$$p_y = \frac{a_{21}(g_X - c_X) + a_{22}(g_Y - c_Y) + a_{23}(g_Z - c_Z)}{a_{31}(g_X - c_X) + a_{32}(g_Y - c_Y) + a_{33}(g_Z - c_Z)} \qquad (21b)$$

where (g_X, g_Y, g_Z) are the coordinates of the ground feature G and (c_X, c_Y, c_Z) are the camera location coordinates in the ground coordinate system (G_X, G_Y, G_Z). The parameters $a_{11}, a_{12}, \ldots, a_{33}$ depend on the camera orientation angles κ, ϕ, ω which orient the image coordinate system (P_X, P_Y, P_Z) with the ground coordinate system (G_X, G_Y, G_Z).

If the angles κ, ϕ, ω are zero, then the imagery is a rectified imagery of the ground at scale of f/H. The ground is not a plane but it is represented on a map as a plane. Thus ground point G is represented as G' on a map. The image of G and G' will not coincide on the image due to the difference in height and this image separation is known as height distortion. A rectified imagery in which height distortion is eliminated is called an orthophoto. Unlike a map, which shows only

Figure 32   Soft photogrammetry system

selected ground features, the orthophoto shows all the features which are sensitive to sensors.

In practice, (p_x, p_y) can be obtained from the pixel location or by direct measurement using special devices known as comparators. The objective of photogrammetry is to determine (g_X, g_Y, g_Z) from (p_x, p_y). Since one image generates two equations with three unknowns, provided $c_X, c_Y, c_Z, \kappa, \phi, \omega$ are known for a pair of imagery (known as conjugate points p, p') we can generate four equations from which g_X, g_Y, g_Z of G are determined.

If the two eyes of a person view the two images, the person gets a stereoscopic view of the feature. Analytical and analogue instruments have been developed to view the imagery in stereo and either plot maps with contours or obtain the digital coordinates of the ground features known as the DTM. Figure 32 shows the latest developments in photogrammetry known as soft photogrammetry in which digital images are used to view stereoscopically and collect DTM using a personal computer (Jeyapalan et al. 1998).

REMOTE SENSING AND GIS

GIS is the science of spatial analysis of information about features on the surface of the Earth. Typically, GIS consists of digital map of the features, database with information on all features, and software to spatially analyze the information and display them in a map, graphically or in tabular form.

Remote sensing is the science of identifying features from the imagery by a detector. Remote sensing compliments GIS. Figure 33 shows the different types of soils identified on an aerial photograph by photo interpretation. In order for this

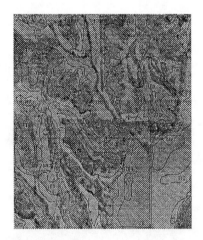

Figure 33   Soil map by photo interpretation

Map from Three Models in
HWY218 Project

By Wu Yao

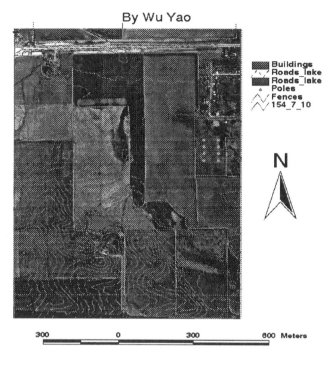

Buildings
Roads_lake
Roads_lake
Poles
Fences
154_7_10

N

300   0   300   600 Meters

Figure 34   Orthophoto with contours

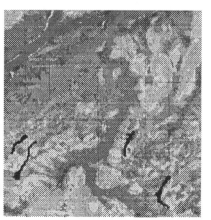

Figure 35   MSS imagery from
Landsat (Drury 1990)

Figure 36   TM imagery from Landsat
(Drury 1990)

identified soil photo to be useful in GIS, it has to be geo-referenced in a spatial dataset in the GIS. The geo-referencing is done by identifying 'tic' marks or reference marks on the soil photo map for which spatial coordinates are available in the spatial dataset. The photo coordinates can then be transferred to the global or ground coordinate system using

$$G_X = a\,p_x + b\,p_y + c \qquad (22a)$$
$$G_Y = d\,p_x + e\,p_y + f \qquad (22b)$$

where a, b, c, d, e, and f are the transformation parameters which can be determined using the 'tic' marks. GIS can spatially analyze the soil types and their locations.

Figure 34 shows an orthophoto with contours and topographic features produced using soft photogammetry (Jeyapalan et al. 1998). It can be geo-referenced to the dataset in GIS and can be used to update the base maps or used with other data for spatial analysis. The scale of the aerial photographs depends on the contour interval. Strips of aerial photographs are taken to stereo cover the required area with sufficient overlap between strips. Each photograph is georeferenced using minimum number of ground control points and the photogrammetric aerial triangulation process.

Figure 35 shows an 18-km width of the MSS imagery from Landsat (Drury 1990). MSS imagery has a swath width of 185 km taken on a Sun Synchronous Satellite orbiting about 700 km above the Earth on near polar orbit. The imagery has 2340 scan lines with 3240 pixels per scan line and it can be geo-referenced, image-analyzed, and used in GIS. Landsat imageries are made in frames covering an area of 185×185km. These frames are corrected for geometric and radiometric distortions before distribution. The user can geo-reference them by identifying ground control points on them whose locations are determined either by using existing maps or by using ground survey methods such as GPS. Figure 36 shows the TM imagery from Landsat that is similarly corrected, geo-referenced, and used in GIS (Drury 1990).

CONCLUSIONS

Remote sensing is the science of identifying features from imagery using a detector mounted on a remote platform. Typically, airplanes and satellites are used as platforms in remote sensing. Imagery obtained from aerial cameras, return beam videocon cameras, multi spectral scanners, and synthetic aperture radar can be used in identifying features for GIS applications. These digital images obtained from detectors are analyzed by computer software using image enhancement and maximum likelihood classifier. In GIS the location of features identified by remote sensing is done on the base map using either photogrammetric principles or geo-referencing to ground points with known locations. Research on developing 3D geographic information systems and virtual reality using remote sensing and soft photogrammetry is currently being conducted.

ACKNOWLEDGMENTS

The author wishes to thank Dr. Len Wilson, Professor of Aerospace Engineering, Iowa state University, for reviewing this chapter. Thanks are also due to Mr. Dwipen Bhagawati, and Mr. Ruifeng Xi, graduate students in Geometronics at ISU for their assistance in preparing this chapter.

REFERENCES

American Society of Photogrammetry & Remote Sensing. (1975). *Manual of remote sensing.* Vol. I & II, Falls Church, VA.

Drury, S.A. (1990). *A guide to remote sensing interpreting images of the Earth.* Oxford Science Publications, New York, N.Y.

Jeyapalan, K. et al. (1998). *Soft photogrammetry for highway engineering.* Engineering Research Institute, Iowa State University, Ames, IA.

Lillesand, T.M. and Keifer, R.W. (1994). *Remote sensing and image interpretation.* 3rd ed., John Wiley & Sons, New York, N.Y.

Slater, P.N. (1980). *Remote sensing.* Addison-Wesley Pub Co., Reading , MA.

CHAPTER 4

Trends in Spatial Databases

Siva Ravada and Jayant Sharma

Spatial databases have been an active area of research for over a decade, addressing the growing data management and analysis needs of spatial applications such as geographic information systems (GIS). This research has produced spatial data types and operators, spatial query languages and processing techniques, and spatial indexing and clustering techniques. In addition, this research also resulted in several extensions to the traditional relational database systems like extensible indexing and extensible optimizers. The objective of this paper is to identify some of the research issues and recent accomplishments in spatial database systems.

INTRODUCTION

Spatial database management systems aim to make spatial data management easier and more natural to users or applications such as urban planning, utilities, transportation, and remote sensing, as discussed in Chrisman (1997). Even though traditional database technology has been evolving for the last thirty years, managing spatial data with database system poses many challenges. Databases are traditionally used in business and administrative applications. In such applications the common data types encountered are integer, float, character, monetary-unit, and date. And the types of operations performed on these data types are simple arithmetic and logical operations like addition, subtraction, less than, greater than, etc. This limited set of data types and operations makes the modeling of real-world spatial applications extremely difficult. Hence, recent research in database systems has focussed on efficiently storing and managing complex data like spatial data. In this paper, we discuss how these new relational databases can be used to solve the problems posed by spatial data management.

A common example of spatial data can be seen in a road map. A road map is a two-dimensional object that contains points, lines, and polygons that can represent cities, roads, and political boundaries such as states or provinces. A road map is a visualization of geographic information. The locations of cities, roads, and political boundaries that exist on the surface of the Earth are projected onto a two-dimensional display or piece of paper, preserving the relative positions and relative distances of the rendered objects. The data that indicate the Earth location (latitude and longitude, or height and depth) of these rendered objects are the spatial data. When the map is rendered, this spatial data is used to project the locations of the objects on a two-dimensional piece of paper. Geographic information system (GIS) is often used to store, retrieve, and render this Earth-relative spatial data. Other types of spatial data include data from computer-aided design (CAD) and computer-aided manufacturing (CAM) systems. Instead of operating on objects on a geographic scale, CAD/CAM systems work on a smaller scale, such as automobile engine or printed circuit boards. Figure 1 shows different sources of spatial data that are widely used today.

The differences among these systems are only in the scale of the data, not its complexity. They might all actually involve the same number of data points. On a geographic scale, the location of a bridge can vary by a few tenths of an inch without causing any noticeable problems to the road builders. Whereas, if the diameters of an engine's pistons are off by a few tenths of an inch, the engine will not run. A printed circuit board is likely to have many thousands of objects etched on its surface that are no bigger than the smallest detail shown on a road builder's blueprints.

All these applications store, retrieve, update, or query some collection of features that have both non-spatial and spatial attributes. Examples of non-spatial attributes are name, soil type, landuse classification, and part number. The spatial attribute is a coordinate geometry, or vector-based representation of the shape of the feature. The spatial attribute, referred to as the geometry, is an ordered sequence of vertices that are connected by straight-line segments or arcs. The semantics of the geometry is determined by its type, that may be a point, line string, or polygon.

What are Spatial Databases?

GIS applications today usually store spatial data and non-spatial or attribute data separately. These systems store spatial data describing the spatial properties of objects in files managed by a file management system. GIS applications then store the attribute data of these objects in a commercial database (like a Relational Database). The reason for this split-data design is partly historical, partly

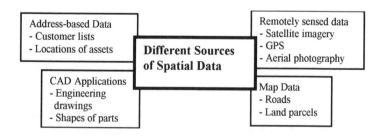

Figure 1 Spatial data comes from different sources

performance, and partly lack of support in the databases for spatial data. This split-data model has several drawbacks, as it is difficult to maintain data integrity between spatial data and attribute data because the two data types are not managed by the same database engine. Programmers must deal with two data streams, one file based and one database-based. The ideal solution is an information infrastructure that includes a single database system for managing spatial data, with a data structure that is independent of the application. In this chapter, we refer to the databases that deal with the spatial and non-spatial attributes in a uniform way as spatial databases.

There are several benefits to managing the spatial and attribute data in a single database. The key benefits of this approach to spatial data management include:

- better data management for spatial data. Traditional GIS users gain access to complete spatial information system based on industry standards with an open interface to their data; that is the Structured Query Language (SQL).
- spatial data is now stored in enterprise-wide DBMS, making it possible to spatially enable many enterprise applications.
- reducing complexity of system management by eliminating the hybrid architecture of GIS data models.
- allowing for seamless integration of MIS and GIS data stores, thus delivering applications that meet the increasingly demanding analysis and reporting needs of a growing geospatial user community.

SPATIAL DATA AND ORDBMS

Traditionally, database management systems are broadly divided into two categories: Relational and Object oriented. Recently, a third type of database systems has become prominent. These systems combine the best of both the relational and object oriented databases (Stonebreaker 1970). These are called the object-relational database management systems (ORDBMS). These systems are developed as an extension to the relational database management systems (RDBMS). In this chapter, we mainly focus on these Object-Relational database systems. First we introduce the main concepts in relational databases.

Relational Database Management Systems

Database management systems have evolved from file management systems that provided simple data-independent routines for sorting, maintenance, and report generation. In these file-based systems, each user defines and maintains the files needed for a specific application. If two users need to manage the same data for two different applications, the users would keep two separate copies of the same data. This makes it hard to share data between two different applications. In the database approach, a single copy of the data is managed and maintained by the database system. And the users can share the same data through open and published interfaces. The initial database systems are built using the relational data model that is introduced by Codd (1970). The relational data model is based on a single logical construct - the relation. A relation can be conceptualized as a table with rows and columns. A tuple is an instance of a thing or entity represented by the table, and the rows of the tuple are instances of the specific data type represented by the column. A common cited example of a table is an employee in a company (Table 1). The rows of this table are instances of the different employees in the company, and the columns represent aspects of the employee such as employee name, number, job title, manager, and department number. Another table or relation called Departments can contain information regarding different departments in a company (Table 2).

The hallmark of the relational model is its simplicity and elegance. It is grounded in an established mathematical theory of relational algebra and first-order predicate calculus. The relational algebra itself consists of a few basic operations: set operations (union, difference, Cartesian product) and relational operations (selection, projection, join). A selection operation is used to select a subset of tuples in a relation (table) that satisfies a condition. For example, one can select all employees who work in a specified department. A projection operation selects certain specified columns from a table. For example, from the Employees table one can select all employee numbers while ignoring other columns like manager, department number, etc. The join operation is the equivalence of a Cartesian product between two tables. For example, the query "Find all employees and their corresponding work locations" will join the Employees table and Departments table based on the DEPTNO column in both tables.

Table 1   Employees data

Table EMPLOYEES				
EMPNO	ENAME	JOB	MGR	DEPTNO
7329	Smith	CEO		20
7499	Allen	VP	7329	30
7521	Ward	Manager	7499	30
7566	Jones	Clerk	7521	30

Table 2   Departments data

Table DEPARTMENTS		
DEPTNO	DNAME	LOCATION
20	Research	Dallas
30	Sales	New York
40	Marketing	Boston

The other advantage of the relational model is that there is a powerful, declarative, and widely used language associated with it: SQL. SQL is the primary interface used to store, retrieve, and manage data in a relational database system. SQL is also used to query the data in the database. For example, SQL can be used to find all the employees who have the same last name. The main feature of SQL is that it is a declarative language, i.e. the users only need to specify what the result is, leaving the actual decisions and logic on how to execute the query to the database system. This makes it easy for different database vendors to provide the same SQL interface to the data, but still have different implementations of the database system. For example, to find all the employees whose department is located in Dallas, the user specifies the query in SQL as:

```
SELECT ENAME, LOCATION
FROM EMPLOYEES, DEPARTMENTS
WHERE LOCATION='DALLAS'
    AND EMPLOYEES.DEPTNO=DEPARTMENTS.DEPTNO;
```

Here the user does not specify in which order the tables are accessed, or whether an index is used, etc. It is up to the database to choose an execution model that it sees fit to execute this query. The result of the query will be the same in any database system as long as they have the same tables and data in them. It is this declarative nature of SQL that makes it an open and standard language.

In addition to the query language, relational databases also provide indexes on the data. Indexes are optional structures associated with tables in a database. The indexes can help speed SQL statement execution on a table. An index is merely a fast access path to the data in the database and it only affects the speed of query execution. And indexes are the primary means of reducing disk I/O during query execution.

The commercial success and wide acceptance of the relational DBMS stems from two ambitious database systems implementation projects: System/R at the IBM San Jose Research Center and INGRES at UC Berkeley. Both of these projects were critical in pushing the relational model to an industrial grade of reliability. In the early 1980s, IBM developed a commercial relational database product called SQL/DS. In 1983, IBM marketed DB2 relational database products for the MVS platforms. DB2 is significant because many relational database vendors attempt to be compatible with DB2's dialect of SQL (The American National Standards Institute, ANSI, proposal was ratified in 1986 and is based on the DB2 dialect.)

Challenges of Spatial Databases

Conventional relational databases often do not have the technology required to handle spatial data. Unlike the traditional applications of databases, spatial applications require the databases to understand more complex data types like points, lines, and polygons. Also the operations on these types are complex when compared to the operations on simple types. Hence we need new technology to handle spatial data. Egenhofer (1993) has identified four main properties of the spatial data that sets them apart from the traditional relational data.

Geometry

Geometry is a main property in any kind of spatial data. Geometry deals with the mathematical properties of an object. These properties include measurement (metric), relationships of points, lines, angles, surfaces, and solids (topology) and order. A simple geometry is usually constructed from the geometric primitive such as points, lines, and areas. Complex geometries are constructed from collections of simple geometries. In addition, there are a number of geometric relationships between two geometries that are very important for dealing with spatial data. For example, a connectivity relation describes how two geometries are connected (e.g. on a road map, how one intersection is connected to another intersection). Metric relationships deal with the distances between two geometries. For example, what are all the cities located within 10 miles of a given road? Geometry is usually represented using a vector data model (where each geometry consists of a set of points) or a raster data model (where each geometry is an image).

Distribution of Objects in Space

Spatial objects are usually very irregularly distributed in space. Consider the case where we model all the cities in the United States as spatial objects (points). Then the distribution of cities on the East Coast is very dense whereas the distribution of cities in the Arizona, Nevada areas is very sparse. In addition, different spatial objects have largely varying extents. For example, if we look at the road network model which models roads with lines and cities with polygons, we see that there are some very large objects in the model (large road like I95) and small objects (like a small city Nashua, NH).

Temporal Changes

Spatial data often have a temporal property associated with them. A simple example is a navigation system that helps travelers find directions from place A to place B in a major city like Minneapolis. If there is an accident on a specific road and that road is temporarily closed for traffic, the system has to adapt to this new data and find a suitable path from point A to point B. When the road clears up for traffic, this new information has to be taken into account in the path computations.

Data Volume

Several GIS applications deal with very large databases of the order of terabytes. For example, remote sensing applications gather terabytes of data from satellites every day. Similarly data warehousing applications and NASA's Earth Observation System are other examples of systems with terabytes of spatial data.

Requirements of Spatial Database System

Any database system that attempts to deal with spatial applications has to provide the following features:

1. A set of spatial data types to represent the primitive spatial data types (point, line, area), complex spatial data types (polygons with holes) and operations on these data types like intersection, distance, etc.
2. The spatial types and operations on top of them should be part of the standard query language that is used to access and manipulate non-spatial data in the system. For example, in case of relational database systems, SQL should be extended to support spatial types and operations.
3. The system should also provide performance enhancements like indexes to process spatial queries (range and join queries), parallel processing, etc., which are available for non-spatial data.

A Solution: Object-Relational Databases

Object-relational database management systems are an attempt to incorporate object-oriented capabilities to a database environment. The new constructs added to the core functionality of traditional relational databases include abstract data types, object identity, and the ability to create operations or procedures through the database programming interface to work on these objects. An example of a project that proposed object-relational (or extended-relational) systems is POSTGRES. Commercial products include the Universal Servers from Oracle, Informix, and IBM. More interestingly, the ANSI standardization committee for the database data language has proposed several extensions to the SQL3 standard that incorporate object-oriented features into the SQL language (Gardels 1997, OGC 1998). Any Spatial database system should address the following five main areas to support spatial applications: (i) classification of space (ii) data model, (iii) query language, (iv) query processing, and (v) data organization and indexing.

Classification of Space

Spatial data consists of objects in space and a description of the space. Some applications require that the system be able to represent different objects in space each of which has a geometry associated with it. This requirement is closer to the vector representation of spatial data. Some applications require that the system be able to describe the space; that is, describe every point in space. This requirement is closer to the raster representation of spatial data. In this chapter, we focus only on the vector models for spatial data.

For modeling different objects in space, the basic elements are point, line, and area. A point represents an object that only has its location in space (X,Y or X,Y,Z) as the spatial attribute. A point does not have an extent or an area, and it can be used to model a city or a building in a large-scale map. A line represents an object that has location attributes along with an extent. A line does not have an area and in the context of spatial databases, a line is always assumed to mean a sequence of connected line-segments in space. A line can be used to model roads, rivers, and utility lines. A region (or a polygon) has location attributes along with extent and an area. Here a region can be a polygon with holes as long as there is only one contiguous area associated with it. Regions can be used to model county boundaries, state boundaries, etc.

Space is also a framework to formalize specific relationships among a set of spatial objects. Depending on the relationships of interest, different models of space such as topological space, network space, and metric space can be used. Topological space uses the basic notion of a neighborhood and points to formalize relationships that are invariant under elastic deformation. Topological relationships include closed, within, connected, overlaps, etc. Network space deals with such relationships as shortest paths and connectivity. Metric spaces formalize the distance relationships using positive symmetric functions that obey the triangle inequality.

Data Model

In traditional database applications the data types of the attributes are limited. These data types consist of integers, floats, character strings, and dates. Furthermore there is no provision for user defined data types. This constraint limits the users in properly mapping their spatial data types to the types supported by the database. What is required is the ability for the users to define their own data types that will naturally map the spatial data types into the database.

Object relational databases provide a higher level of abstraction for spatial data by incorporating concepts closer to human's perception of space. This is accomplished by incorporating the object-oriented concept of user-defined abstract data types (ADTs). An ADT is a user-defined atomic type and its associated functions. For example, if we have land parcels stored in a database then an ADT would be a combination of the 'atomic type' polygon and some associated function, say adjacent, that may be applied to land parcels to determine if they are adjacent. The label 'abstract' is applied because the database is unaware about the meaning of the user-defined data type and the implementation details of the associated methods. All it needs to know are the available methods and their input/output types. This encapsulation allows for a seamless integration of predefined and user-defined data types.

Query Language

A query language provides the means to access and manipulate data in the database. The query language should have enough constructs built into it to express a wide variety of data types. At the same time it must be intuitive and easy to use. A popular query language for relational databases is SQL. It is known that the traditional SQL is inadequate to express typical spatial queries. This has prompted various efforts to extend the capability of SQL with spatial-friendly constructs. At the same time, the standards committee is currently working on a draft to update and make it compatible with the generic functionality offered by object-relational database management systems. The Open GIS Consortium led by important GIS and database companies has come out with their own proposal to include GIS capabilities in SQL (OGC 1998). The specification is described by a standard set of Geometry Types based on the OGC geometry model, together with functions for these types. Common spatial operators like adjacent, overlap, and inside, and functions like buffer, perimeter, and area have been included in the specification.

Spatial Query Processing

Spatial queries are often processed using filter and refine techniques; for a survey of different spatial query processing techniques see Gueting (1994). In the first filter step, some approximate representation of a spatial object is used to determine a set of candidate objects that are likely to satisfy the given spatial query. Common approximations used in spatial databases are minimum orthogonal bounding box (Figure 2) and multiple bounding boxes. Then the candidate set is further examined using the exact representations of the objects to find the actual set of objects that satisfy the given query. The approximations are chosen such that if the approximations of objects A and B do not satisfy a relationship, then that relationship cannot be satisfied between the objects A and B.

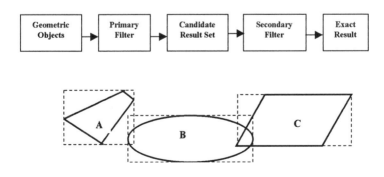

Figure 2   Spatial query model

Figure 2 illustrates this filter and refine strategy. Figure 2 shows three spatial objects A, B, and C along with their minimum orthogonal bounding rectangles (MOBR). Here the MOBR of A and B intersect, but the actual geometries of A and B do not intersect. Conversely, if two MOBR do not intersect, then the actual geometries corresponding to these two MOBR also do not intersect. And this technique is used in most of the spatial query models.

There are several advantages to this filter and refine strategy. First, spatial objects tend to be very large and hence will put considerable strain on a database system to retrieve the whole object into the main memory. Compared to this, an approximate representation of a spatial object will take considerably less processing time to load into memory. Second, computations on spatial objects tend to be very complex and time consuming. And the more complex the object is, the more is the processing power required to compute the spatial relationships. Compared to this, the computations using approximate objects tends to be very fast and hence can save large amount of processing power.

Like any other queries in a relational database, spatial queries are classified into two categories: (i) selection queries and (ii) join queries. Selection queries in spatial databases are also known as window queries as the selection condition is often expressed as a window object. Window queries take a spatial object (called the window object) and seek spatial objects from a table that satisfies a binary relation with the window object. For example, the query "Find all the roads in the City of Minneapolis which overlap the park Minnetonka" is a window query. Here the object representing the park Minnetonka is the window object. A join query seeks all the object pairs from two tables that satisfy a given relation. For example, the query "Find all the roads and parks which overlap in the city of Minnesota" is a join query.

Spatial Indexing

The purpose of a spatial index is to facilitate spatial selection. That is, in response to a query, the spatial index will only search through a subset of objects embedded in the space to retrieve the query result set. There are essentially two ways of providing a spatial index. First, dedicated external spatial data structures are added to the system, offering for spatial what a B-tree does for standard relational attributes. Second, spatial objects are mapped into a one-dimensional (1-D) space so that they can be stored within a standard 1-D index such as the B-tree index. Apart from spatial selection, spatial indexing also supports other operations such as spatial join, nearest neighbor, etc.

A fundamental idea for spatial indexing is the use of approximations. This allows index structures to manage an object in terms of one or more spatial objects, which are much simpler geometric objects than the object itself. The prime example is the bounding box (the smallest orthogonal rectangle enclosing the object). Another method, called grid approximation, divides the space into cells by a regular grid, and the object is represented by the set of cells that it intersects. The use of approximations leads to a filter and refine strategy for spatial query processing. First, based on the approximations, a filtering step is executed; it returns a set of candidates that is a super set of the objects fulfilling a spatial predicate. Second, the result is refined by checking the exact geometry of each candidate (or a pair of candidates in the case of a spatial join).

COMMERCIAL SPATIAL DATABASE SYSTEMS

Initial spatial systems had applications that ran on top of a file system. All the spatial data is stored in flat files and each GIS application is responsible for maintaining and managing the corresponding spatial data. The corresponding attribute data, on the other hand, is stored in a database system (relational in most cases). The application also had to maintain the spatial index on top of the file system. These spatial systems are adequate for small-scale applications that only deal with small amounts of data. Since the data is managed by the applications on top of a file system, these systems have several limitations. Applications could not combine queries that deal with spatial and non-spatial attributes, as these data are stored in two different places. It also made sharing of data across multiple applications very difficult.

This lack of flexibility in accessing the spatial data resulted in the recent systems where the database systems started managing the spatial data in the database. Initial versions of these spatial database systems have the application store the spatial data in the database in the form of a lob (large object). The database itself does not understand the content of the lob, nor does it know how to interpret the spatial attributes. Spatial applications still have to interpret the spatial attributes of the data, but the data itself is stored and managed by the database. These systems have the advantage that the spatial data is now centrally located in the database along with the other attribute data. And hence the database can take care of the data management issues like backup, recovery, and integrity. Furthermore, several applications can share the same data store. ESRI's SDE is an example of one such commercial system. However, these systems still lack flexibility and do not completely integrate the spatial data with other attribute data. The GIS application still uses a proprietary interface to read and understand the spatial data whereas the attribute data can be accessed via a standard interface like SQL. The database system still does not understand the spatial data, which means that the queries cannot be optimized by taking into account the spatial properties of the query. The next generation spatial database systems addressed most of these drawbacks.

Object-relational database systems provide the GIS applications the ability to completely integrate spatial data with attribute data in the database system. The ORDBMS provides SQL extensions so that spatial data can be accessed and managed like any other attribute data. They also provide user defined indexes and functions that let the database understand the spatial operations during the query optimization phase. Oracle's Spatial, Informix 2D Datablade, and IBM's DB2 Spatial Extender (Chamberlin 1997) are examples of commercial systems that

71

provide this functionality. In this section, we review how current commercial database systems handle spatial data and we use Oracle Spatial database as the main reference.

Oracle Spatial

Oracle8i Spatial provides a completely open and standards-based architecture for the management of spatial data within a database management system. Users can use the same query language SQL for accessing spatial data as well as other non-spatial data in the database. The functionality provided by Oracle8i Spatial is integrated within the Oracle database server. Users of spatial data gain access to standard Oracle8i features, such as a flexible client/server architecture, object capabilities, and robust data management utilities, ensuring data integrity, recovery, and security features that are difficult to obtain with other architectures. Oracle8i Spatial enables organizations to merge GIS and MIS (management information system) data stores and implement unified data management architecture for all data across the enterprise. The Oracle8i Spatial provides a scalable, integrated solution for managing structured and spatial data inside the Oracle server. This architecture enables users to exploit the power of today's multiprocessor hardware; an essential requirement for enterprise scale spatial solutions.

Spatial Data Modeling

Spatial cartridge supports three geometric primitive types and geometries composed of collections of these types. These primitive types are: (1) point and

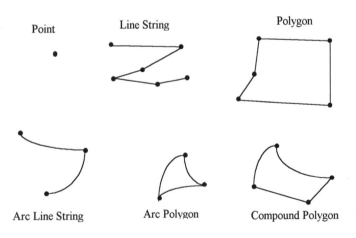

Figure 3   Geometric shapes in Oracle Spatial

point cluster, (2) line string, and (3) and N-point polygon, where all these primitive types are in two-dimensions. A 2-D point is an element composed of two coordinates, X and Y. Line strings are composed of one or more pairs of points that define line segments. Any two points in the line segment can be connected either by a straight line or a circular arc. That means line strings can be composed of straight line segments, arc segments, or a mixture of both. Polygons are composed of connected line strings that form a closed rind and the interior of the polygon is implied. Since polygons are composed of line strings this implies that a polygon can have some edges as straight lines and some edges as circular arcs. Figure 3 shows some example geometries that can be represented in Oracle Spatial. The spatial data model is a hierarchical structure consisting of elements, geometries, and layers. Spatial layers are composed of geometries that are in turn composed of elements.

Figure 4 describes the hierarchical model of spatial data in Oracle Spatial. An element is the basic building block of a geometry. For example, elements might model star constellations (point cluster), roads (line strings), and county boundaries (polygons). In case of polygons with holes (an island within an island) the exterior ring and the interior ring of the polygon are considered as two distinct elements that together make up a complex polygon. A geometry is the representation of a users spatial feature, modeled as an ordered set of primitive elements. A geometry can consist of a single element, or a homogenous or heterogeneous collection of primitive elements. A multi-polygon, such as one used to represent a set of islands is a homogenous collection. A heterogeneous collection is one in which the elements are of different types. A layer is a heterogeneous collection of geometries that share the same set of attributes. For

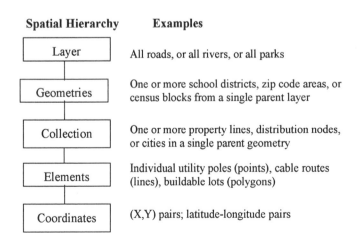

Figure 4   Data model hierarchy

example, one layer in GIS might include topographical features, while another describes population density, and a third describes a network of roads and bridges in an area.

Operations on Spatial Data Types

The binary topological relationships between two spatial objects A and B in the Euclidean space is based on how the two objects A and B interact with respect to their interior, boundary, and exterior. This is called the 9-intersection model for the topological relationships between two objects. This can be concisely represented using a 3x3 matrix. From this matrix, one can theoretically distinguish $2^9 = 512$ binary relationships between A and B. In case of two-dimensional objects, only eight relations can be realized which provide mutually exclusive and complete coverage for A and B. These relationships are: contains, covered by, covers, disjoint, equal, inside, overlap, and touch. Figure 5 illustrates these relationships between two spatial objects, A and B. Oracle Spatial supports this 9-intersection model for determining the topological relationships between two objects. In addition, other relationships can be derived as a combination of the above 8 relations. For example, OVERLAPBDYDISJOINT can be defined as the relation where the objects overlap but the boundaries are disjoint.

Distance functions are also one of the widely used binary relationships. Distance functions relate two objects depending on the distance between the two objects. A common distance function used in spatial application is a within distance function. This function returns true if two objects are within a specified distance of each other. Oracle Spatial provides this within-distance function. In addition, this system provides a set of theoretical operations like UNION, INTERSECTION, DIFFERENCE, and SYMMETRIC-DIFFERENCE. For example, given two spatial objects A and B, the database system can compute and return a new object C which is the UNION of A and B.

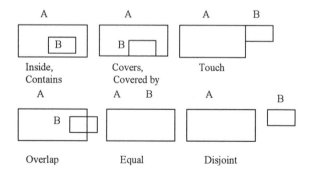

Figure 5 Spatial relationships in Oracle Spatial

SQL Support for Spatial Data

Query language is the principal interface to the data stored in a relational database system. Traditional SQL has been extended in ORDBMS to be able to support new data types. In the case of Oracle Spatial, SQL is extended in two ways: SQL can be used to define and create objects of spatial types. SQL can also be used to insert, delete, and update spatial types in addition to being able to query the spatial data with the help of spatial functions. For example, all the parks in a city that overlap the rivers in that city can be found using the SQL query:

```
SELECT parks.feature
FROM parks, rivers
WHERE sdo_relate(parks.geometry, rivers.geometry, 'OVERLAP') = TRUE;
```

Spatial Indexing Using Linear Quadtrees

One of the key features of Oracle Spatial is the ability to index spatial data. An index is required to be able to efficiently process queries like find objects within a data space that overlap a query area, usually defined by a query polygon (window object) and find pairs of objects from within two data spaces that spatially interact with one another (spatial join).

An index in Oracle8i Spatial is a logical index. The entries in the index are dependent on the location of the geometries in a coordinate space, but the index values are in a different domain. Index entries take on values from a linearly ordered integer domain while coordinates for a geometry may be pairs of integer, floating, or double-precision numbers. Oracle Spatial uses a linear quadtree-based indexing scheme, also known as z-ordering, that maps geometric objects to a set of numbered tiles. A tile in 2-D space is a box whose edges are orthogonal to the two coordinate axes. The coordinate space where all the geometries are located is decomposed in a regular hierarchical manner. The range of coordinates (the coordinate space) is viewed as a rectangle. At the first level of the decomposition, this rectangle is divided in half in each of the coordinate directions, forming four sub tiles called quads. At each subsequent level, each quad is divided in half in each coordinate direction forming four sub tiles. As the name implies, the four-leaf structure of a quadtree can be used to build an index tree. This process continues until some termination criteria, such as the minimum size of the tiles, is met. These tiles can be linearly ordered using some ordering scheme like the z-ordering.

The spatial index consists of at least two columns: a foreign key to the object being indexed and a cell key to some cover of space. It is assumed that the object is covered by the cells to which it is associated. We will further assume that no cell intersects the object unless it is in the index or is a sub cell of a cell in the index. Point data can be very well indexed by a recursive decomposition of space. Spatial object with extent, such as area or line features, creates a problem for this sort of index, because they are highly likely to cross index cell partition boundaries. In Oracle Spatial, each item is allowed multiple entries in the index. This allows one to index features with extent by covering them with the decomposition tiles from a hierarchical decomposition.

Query Processing

For a window query, the window object is tiled using the decomposition of the index. Search candidates consist of all leaf nodes of the index that are associated with index keys that overlap any of the tiles of the window. A spatial join is a query that depends upon a spatial relationship between geometric attributes in two columns. One mechanism for the spatial join is to use the first data space to drive a sequence of window queries on the second data space. Another mechanism can be used in case where both the tables are indexed. Each tile associated with the first table would require visiting that same tile (or the surrounding tiles if a within-distance were part of the query) in the second tables index space. This means that the join could be accomplished by a simultaneous traversal of both indexes.

CONCLUSIONS

Recent reports have described the accomplishments of spatial database research and have prioritized the research needs in this area. Gueting (1994) and Worboys (1995) provide a broad survey of spatial database requirements and an overview of research results. Kim et al. (1993) discussed the research needed to improve the performance of spatial databases in the context of object-relational databases. The primary research needs identified were SQL support for spatial types, support for spatial indexing methods, development of cost models for query processing, and the development of new spatial join algorithms. Many of the research needs identified in this research have since been addressed.

Several commercial spatial database systems exist today, that provide solutions to many of the problems identified by such research work. Current spatial database systems provide support to store and retrieve geometric properties of spatial features. They also provide indexing support and operational support like topological and distance functions. The main deficiency in current systems is that they lack proper support for managing different spatial reference and projection systems. They also lack support for applications like transportation systems that require linear referencing capability. Support is also required for applications with spatio-temporal data and geo-referenced imagery. Several database products have already started addressing these issues and many other open problems in spatial databases.

REFERENCES

Codd, E. (1970). "A relational model for large shared data banks." *Communications of the ACM*, 13(6), 13-27.

Chamberlin, D. (1997). *Using the new Db2: IBM's object relational system.* Morgan Kaufmann, San Fransisco, CA.

Chrisman, N. (1997). *Exploring geographic information systems.* John Wiley & Sons, New York, N.Y.

Defazio, S. et al. (1995). "Integrating IR and RDBMS using cooperative indexing." *Proc., SIGIR,* Seattle, Washington.

Egenhofer, M.J. (1993). "What's special about spatial? Database requirements for vehicle navigation in geographic space." *Proc., ACM SIGMOD Conf.,* Washington, D.C., 398-402.

Gardels, K. (1997). "Open GIS and on-line environment libraries." *ACM SIGMOD Record,* 26(1), 32-38.

Guting, R.H. (1994). "An introduction to spatial database systems." *VLDB J.,* 3, 357-399.

Kim, W., Garza, J., and Kesin, A. (1993). "Spatial data management in database systems." *3rd Intl. Symposium on Advances in Spatial Databases,* Singapore, 1-13.

Open GIS Consortium. (1998). Oracle8i spatial cartridge user's guide and reference, Release 8i. Internet: http://www.opengis.com.

Stonebreaker, M. and Kennitz, G. (1993). "POSTGRES next-generation database management system." *Communications of the ACM,* 34(10), 78-92.

Stonebreaker, M. and Moore, D. (1997). *Object relational DBMSs: the next great wave.* Morgan Kaufmann, San Fransisco, CA.

Worboys, M.F. (1995) *GIS: a computing perspective.* Taylor and Francis, Bristol, PA.

CHAPTER 5

Implementation of Linear Referencing Systems in GIS

Wende O'Neill and Elizabeth A. Harper

For decades, state Departments of Transportation (DOTs) have been collecting and using information about the highway systems based on linear location references. Linear referencing systems (LRS) have been developed to manage and maintain linear location reference data. Recently, geographic information systems (GIS) have been proposed as a technology to facilitate integration, management, and analysis of data used by DOTs. Unfortunately, implementation of linear referencing systems within GIS has not been as straightforward as hoped. This chapter presents an overview of linear referencing systems with a discussion of the advantages and disadvantages of various linear referencing methods, followed by an overview of GIS for transportation (GIS-T). The need for linear referencing in GIS-T is discussed, with emphasis on the spatial data transfer standard (SDTS) transportation network profile, intelligent transportation systems, and dynamic segmentation.

INTRODUCTION

Linear referencing systems (LRS) have been used for decades by state Departments of Transportation (DOTs) to record and manage the location information associated with the characteristics and condition of transportation facilities. For example, the location of permanent loop detectors for traffic counts may be identified by route names and nearest mile points and offset along routes. Most DOTs use several methods (both linear and geographic) for referencing location. The common methods include identifying location based on a mile point or reference post, using street names at intersections, street addresses, and control section and link identification numbers, and recording the spatial coordinate pair (such as latitude and longitude, state plane or UTM address).

Recently, many DOTs have implemented geographic information system (GIS) technology to store, manage, manipulate, analyze, and report spatially referenced data used in the agency. The adoption of GIS within an agency has required that the technology support linear referencing methods used in the DOT's legacy databases. A great deal of thought and research has been focused on the issue of how to implement LRS in GIS.

The Bureau of Transportation Statistics (BTS) has compiled and produced a reference tool to support state and local transportation agencies with implementation of LRS in GIS-T. The CD-ROM contains a resource guide that provides an overview on the subject, over 100 scanned documents on models and applications and state reference manuals. It also has a searchable database with links to Internet sites containing more documents on the subject, and an example of a robust implementation of LRS in GIS by the Wisconsin DOT. This chapter presents a synthesis of the topics covered in depth in the Resource Guide on LRS implementation in GIS.

Background

Linear referencing is a means of identifying a location on a linear feature, such as a road or railroad. It provides a means by which people can communicate an understanding of location. An address such as 235 Main Street is a linear reference. Three miles north of the junction of routes 1 and 50 on Rt. 1 is another linear reference.

Many types of linear references have been developed over the years as people have attempted to communicate about roadways and locations on roadways. In early years, there was little need to standardize the means of communicating about location; as long as there was a common understanding of the area, locations could be referenced with common landmarks, such as "the accident occurred down by Earl's Garage." But as the U.S. highway network grew and became recognized as an infrastructure that needed to be inventoried, maintained, and managed, a need to standardize reference locations along the highway grew as well. Linear referencing systems accommodate this need.

A linear referencing system incorporates a linear location reference plus the commonly understood meaning of the references and the field data (markers or house numbers) that allows for actual identification of the reference in the field. A system requires not only a common understanding of how to refer to a location, but also a means to identify the location in the real world. For example, an address is meaningless as a linear location reference if there are no house numbers posted and there is no common means (such as a map) to communicate the location and names of the streets.

Transportation agencies adopt linear referencing systems to communicate with others about events occurring on the highway system. These events may include information about poor pavement condition or the need for roadside mowing, or they may also include incidents such as crashes, breakdowns, or spills. LRS may be used to assign crews to snow plowing routes or to send a highway patrol car out to a vehicle that has broken down. With a linear referencing system in place a motorist on the road can report his/her location or find a location. It allows a motorist or transportation official to answer the questions 'Where am I?' and 'How do I get there?' (Deighton et al. 1994). With the development of geographic information systems, transportation agencies have found another reason for implementing and maintaining a linear referencing system: to visualize event information and perform spatial analysis on maintenance and management data.

COMMON LINEAR REFERENCING METHODS

It is important to distinguish between location referencing methods and location referencing systems. A location reference method (LRM) is a set of procedures used in the field to identify the address of any point. The objectives of linear location reference methods are to designate and record the geographic position of specific locations on a road and use the designations as a key to stored information about the locations (NCHRP 1974). In general, linear location referencing methods use a *name* (e.g. route number), *direction,* and a *distance measurement* (e.g. offset) from a *known location* as an *address* for an object or event. A linear reference system is a set of procedures used in an agency to manage all aspects of linear location referencing. The *system* includes procedures for storing, maintaining, and retrieving location information (Deighton et al. 1994, NCHRP 1997). A location referencing method is the way that the linear location referencing system identifies location.

There are many linear referencing methods that can be incorporated into an agency's linear referencing system, including route mile point, route reference post offset, and street address, among others. In every case, each linear feature (road, railroad, river, etc.) for which a reference is assigned has a primary direction. This may be the direction in which one is traveling when the data are collected. Most states have adopted a common convention regarding the primary direction of their state routes; from south to north and from west to east. It should be noted that direction in LRMs *has nothing to do with the non-topological property of bearing.* Rather, direction in an LRM is the *topologically invariant property of connectedness* (Worboys 1995).

Cities, counties, and townships also have adopted an implicit direction element in their address numbering schemes. For example, in Utah, street numbers increase as distance increases from a central origin (usually State and Center Streets) in each urban area. Exceptions abound for any rule implying direction (for instance, even numbered routes have an eastbound primary direction). However, it is important to note that direction is a fundamental component of any LRM. Each linear referencing method has its strengths and weaknesses. The common methods are described below.

Route Mile Point Method

Overview

The route mile point method uses the measured distance from a given or known point to the referenced location. The beginning point may be the beginning of a route, the point at which the route enters a state, or the point at which the route enters the county or town. The term 'route' in this context means a collection (one or more) of roads that have been grouped for administrative purposes under a common name or identifier. The measured distance, or offset, typically is the accumulated mileage from the known point to the referenced location.

It is important for the field data collector to know the starting point and the primary direction of the route when reporting a location. The collector should also be aware of the precision of the offset needed to report the locations (0.1, 0.01, 0.001 of a mile or kilometer) and the measurement position (along the centerline, the shoulder lane, the median lane, etc.).

Treatment of discontinuous routes in the LRS must also be understood (Figure 1). Some states treat all routes as continuous and record mile point information for each route on a shared alignment. Others identify a primary or key route and only maintain mile points for the key route on a shared roadway section. In this context, a shared alignment is a shared roadway section – it is when one piece of the road gets more than one name (e.g. where I-95 overlaps I-495). A route is discontinuous when the administration assigns a primary name to an overlapping section. So, if SR 91 and SR 89 share the same road alignment for 10 miles and SR 91 is the primary route, SR 89 is discontinuous.

Divided highways, couplets, and route deviations are transportation facilities that deserve special consideration. Most LRM treat a road as a single feature. On these facilities, each section of pavement that accommodates a particular travel direction or pattern should be referenced to clearly indicate where an event occurs. Many states treat divided highways as separate facilities and calibrate the mile points in both directions. However, couplets and route deviations are handled using a variety of approaches.

Advantages and Disadvantages

The mile point method is the fundamental linear referencing method that is used to relate all other linear referencing methods. For example, accumulated mileage to a reference post or point (e.g. bridge abutment) is required to identify

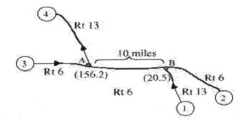

Figure 1   Discontinuous routes

the location of the reference point in the field. An advantage to using the mile point method is that reference posts or signs do not have to be maintained in the field, as discussed later.

However, the mile point method requires that the person in the field go back to the beginning of the route to start measuring distance in order to locate a reference. A person in the field must know where the route begins (state line, county line, or town line) and the primary direction of the linear referencing system. For county or town line-based approaches, some indication of jurisdiction is necessary. For instance, the reference Rt. 95 MP 1.5 actually identifies several locations along Rt. 95 in a county-based method – one in each county that the route passes through. Therefore, a county name, like Fairfax, is needed to differentiate this address from the others.

Another disadvantage of the mile point method is the burden of maintaining a historical record of changes to the linear referencing system. If a roadway is realigned, all of the roadway beyond that point will suddenly have different mile points. The system is burdened with maintaining historical relationships so that two crashes occurring in exactly the same location in different years and reported at mile point 321.3 one year and 322.0 the next year are understood to have the same real-world location. Some DOTs use 'equations' and special identifiers to indicate that some change has occurred and to keep the mile points along a route consistent beyond the point of the change.

Route Reference Post Offset

Overview

The route reference post-offset method uses signs posted in the field to indicate known locations. The difference between mile posts and reference posts is that mileposts are placed at every mile. Some states treat mile posts as reference posts given the potential inaccuracy of their location. Other states post signs with mile point information or an identification number that is related back to a mile point in the LRS on bridges and other prominent features along a road. Events in the field are measured as an offset from those posted references. For example, a reference post numbered 200 might be posted on a bridge on I-93. Moving in the positive direction (that is, in the primary direction), a crash might be reported at I-93 reference post 200 offset 2.1 miles. The reference post-offset method allows data collectors to report a negative offset. The negative offset would be recorded as a distance before a reference post. In the example above, if there are three miles between reference post 200 and 201 then the same accident could be recorded at reference post 201 minus 0.9 miles.

Advantages and Disadvantages

The primary advantage of the reference post-offset method over the mile point method is that it is not necessary to return to the beginning of the route in order to begin measuring the distance along the route to the incident. It is only necessary to go back, or forward, to the nearest reference post. Another advantage is that the reference post need not reflect the exact mileage along the route: it can be an arbitrary, unique number. However, it is easier for the field personnel and the motoring public to understand where they are if the reference posts follow a sequence from the beginning of the route. Reference posts do not need to be evenly spaced along the roadway, but typically reference posts are placed about one mile apart.

When realignment occurs, only that portion of the roadway, which is realigned, is affected; the reference posts beyond the realignment remain in their same relative position. For example, a crash may occur 2.1 miles from reference post 200 on I93. Suppose there is realignment that changes the length of the roadway before reference post 200. In the future, a crash occurring at exactly the same place will still be reported at 2.1 miles from reference post 200. Reference post 200 will be a different distance from the beginning of the route.

A disadvantage of the route reference post-offset method is that all the reference posts must be maintained in the field. It is important that their location be exact. For example, if signs are knocked down by a snow plow during a heavy snow winter, it would be essential that they be replaced at exactly the same location. This will require that the maintenance crew carefully measure the distance between the last remaining upright post and the new replacement posts.

An additional problem is that of maintaining visible reference posts in urban areas where signs are overly abundant. It may not always be possible to put a reference post in the exact location that is optimum for the LRS. Even if it were possible to establish all the necessary reference posts, reference posts in dense urban areas are hard to identify. Some states, like New York, do maintain markers in urban areas. Others, like Wisconsin, use fixed transportation features, like bridges and light posts as reference points in urban areas.

Street Address

Overview

The street address can be used to define where on a linear feature an event has occurred. This can be a particularly useful method in dense urban areas where reference posts are difficult to locate. It is less effective in rural areas where addresses are not necessarily evenly spaced and are hard to identify. Transportation applications using street address-based methods are most appropriate on local roads. Street address methods have been gaining popularity over the last decade due to the efforts related to Enhanced 911 (E-911) programs to assign street addresses to all buildings (commercial and residential) in suburban and rural areas.

Advantages and Disadvantages

The advantages of using a street address spring primarily from the use of the method in an urban area where addresses are fairly easy to identify and report and where house numbers are fairly evenly dispersed along a roadway. The street address has the advantage of being a method that many people understand and can easily report from the field. However, in rural areas addresses are less evenly dispersed and are harder for the field data collector to ascertain. Furthermore, many highways, particularly interstate roads and some limited access roads, do not have postal addresses. The street address method of linear referencing must be used in conjunction with other methods in order to obtain complete coverage of a region.

Street address methods differ from the previous two in that an address refers to a discrete, and potentially discontinuous, space. For instance, vacant lots and agricultural parcels in suburban areas may not be assigned a street address. Street addresses must be evaluated as a discrete numbering system. Interpolation algorithms that work with continuous numbering systems will often generate invalid addresses: a location midway between 246 Main St. and 266 Main St., namely 256 Main St, does not exist in the real world. There may be several addresses per block or a single address associated with an entire block face (for instance, the Nassif Building in Washington D.C. that houses the USDOT consumes an entire block.). Referencing a manhole entrance or crash to the address (400 7th St) may not give a precise or clear description of the location.

Another difference between the street address method and other methods is that direction and offset are implied, not stated or measured. Unlike other methods, separate street addresses exist (conceptually) for both sides of a single transportation facility. Consequently, one can infer from the address, for example, the group of travel lanes in which a crash has occurred. Alternatively, there may be two valid addresses for an event (e.g. 247 Maple St. and 246 Maple St.).

Link-Node Models of Location Referencing

Overview

Earlier work on location referencing describes link (chain) and node as a location reference scheme (Nyerges 1990). It is more of a scheme than a method because of the similarity between link-node and other referencing methods. Basically the link-node method uses node numbers, identified with physical features in the field such as an intersection. Links are the logical connection between nodes. All events that are recorded are measured as an offset from the nearest node number along a link. The field data collector must carry a node map to be able to identify the location of the nearest node and to record the node number. Then the data collector must measure the distance from the node to the event.

As noted above the link-node method or scheme is similar to other referencing methods. For example, it is similar to the reference point method. In both approaches an offset is measured from a known point in the field. That point may be a bridge abutment, an intersection, or any other physical feature. In both cases, a reference such as a map is required for the person in the field to identify the known point from which to calculate the offset. The difference between the reference point method and the link-node scheme lies in the understanding that a link-node system is the framework for tabular representation of a network. Note that in the glossary there are many different definitions of links, nodes, and networks, adding to the confusion when discussing link-node as a linear referencing method.

Other schemes similar to the link-node method include the use of control sections and the 'Document Method II' scheme described in NCHRP synthesis *Highway Location Reference Methods* (NCHRP 1974). Control sections may be represented in a tabular database using this type of scheme, such that a point ID is attached to the beginning and ending of each control section. However, attributes associated with the link between and end points of a control section are homogeneous (Nyerges 1990) so only events like accidents would be reported using an offset from the end point of a control section. The 'Document Method II' described in NCHRP (1974) gives each street a unique five-digit code. Each intersection is defined by a combination of two street codes. Accidents are then referenced using an offset distance from the nearest intersecting curb and the intersection street codes.

Advantages and Disadvantages

When realignment occurs in the network, a link-node method of location referencing does not need to be completely reassigned. The distance between two nodes may change but the distance between all the downstream nodes is unaffected. A link-node method also has the advantage of not needing to maintain reference posts in the field. The reference post is in effect some permanent feature that exists in the field and does not change. Intersections are the primary feature used for nodes; intersections cannot be destroyed by snowplows or obscured by vegetation. Note, however, that not all intersections are necessarily nodes.

The primary disadvantage of the link-node method is that the data collector must carry a node map to be able to identify the node number to report the location of an event in the field. Ordinary operators of vehicles cannot use this method to report the scene of an incident; nor can the DOT communicate to the public information about incidents at a given location. This is likely to be an issue for the implementation of intelligent transportation systems (ITS) which require the ability to communicate information to the vehicle operator about incidents and alternative routes.

GEOGRAPHIC INFORMATION SYSTEMS FOR TRANSPORTATION

Background

Although some form of computer mapping has existed since the 1950s, it was not until the computer revolution of the late 1970s that automated mapping began to see rapid development and broad application. It was not long before the ability to perform automated spatial analysis was identified as a computerized function for geographic data. Analyses that were previously not possible were becoming known and implemented. This was the beginning of geographic information systems.

GIS is used to organize and manipulate data that are collected, maintained, and analyzed by an agency. GIS commonly is used to find solutions to problems with a spatial component. In other words, in these types of problems, *where* something occurs (e.g. bad pavement), in absolute or relative position, is as important as the condition itself.

Data incorporated into GIS are organized by location in the transportation system and on the Earth. GIS allows spatial and descriptive or attribute data management. Spatial data are data about features that take up space or are located on, above, or below the Earth's surface. Spatial data are generally referred to as entities, and are used to determine the physical properties of features, such as size, length, circumference, and area. Attribute data describe characteristics of real world features. For example, a road is a real world feature and its attributes include number of lanes, pavement type, and traffic volume.

The pervasive paradigm in GIS is that entities in the real world can be represented as geometric objects, namely points, lines, and polygons. Part of the data modeling process involves determining how to represent real world entities with geometric properties. For instance, a building may be represented as a polygon showing its footprint on the Earth or as a point indicating its centroid or main entranceway. A city may be represented as a polygon showing its current political boundaries or as a point. A road may be represented as a single centerline or as a polygon defined by its right-of-way or paved surface.

Attributes are associated with a geometric object that represents a real world entity. It is extremely important to consider the attribution of spatial entities while considering the geometric properties by which to represent them. This is particularly true in a transportation GIS where a multitude of attributes are associated with linear objects and the environment surrounding these objects.

Early GIS implementations focused upon the use of polygons and polygon overlay functions for their analysis. Roadways, like other types of line features (rivers, utility lines, etc.), were used to delineate polygons and were not thought of as features for analysis in and of themselves. Even as recently as 1990, Star and Estes (1990) define a *line* as: *"a spatial object, made up of a connected sequence of points. Lines have no width, and thus, a specified location must be on one side of the line or the other, but never on the line itself."*

In the late 1980s and early 1990s transportation professionals began to look at roadways as distinct geographic features which could be analyzed using GIS technology. For all but engineering level scales (e.g. 1:200) roads were represented as single centerline features using a vector data structure. Transportation professionals need to locate events (crashes, pavement conditions, and inventory information) along lines as well as map the coincidence of different roadway characteristics and events (e.g. poor pavement and rate of superelevation on curves). The subsection of GIS that focused on transportation came to be known as GIS-T.

It was immediately apparent that the usual spatial analysis techniques of polygon overlay functions would not meet the needs of spatial analysis of linear features such as transportation systems. There was not a function or tool for the analysis of linear features similar to polygon overlay. Further, an interesting dialog ensued on how best to collect and record roadway data in order to associate attributes of roads with geographic lines or arcs in a vector-based GIS.

Data Representation Methods

Spatial Data Representation

A vector-based geographic database consists of collections of points, lines, and areas or polygons. A point is recorded as a single (x, y) coordinate pair, a line is a series of (x, y) coordinates, and an area is a closed loop of (x, y) coordinate pairs. The 'Topological Model' is the most common method of encoding spatial relationships in vector GIS (Arnoff 1993). The common name is the Arc-Node Model where a node is an intersection where two or more arcs meet.

The Arc-Node model is very similar to the link-node approach used to model transportation flows in travel demand forecasting packages. However, arcs have intermediate, or shape, points defined by (x, y) coordinate pairs that reflect, for example, the path along the ground followed by a road. Also, since the spatial representation of arcs may not be used to model flows, it is not necessary to create nodes only where changes in direction of flow may occur. Finally, since arcs do not necessarily reflect (traffic) flow, it is not necessary to have multiple arcs representing flow direction. In the link-node model for transportation planning applications, each link corresponds to a single flow direction.

Transportation roadway centerline databases (spatial representations) within a GIS typically are created by digitizing along a road feature, collecting a series of (x, y) coordinate pairs (note that it is not necessary to digitize both roadbeds in a divided highway). Some data models capture the digital line as the centerline of the median and describe the arcs in the digital centerline as representing a divided highway so that they can be drawn using parallel line symbols. Others generate a separate digital line at a fixed distance from the centerline of the first digitized roadbed. Only at a very large scale (e.g. 1:1000) are divided highways typically captured as two separate lines. Arcs may begin and end at intersections with other roads, boundaries, and other transportation features, such as railroads, canals, and rivers.

The segmentation scheme (where an arc starts and stops) depends on the spatial database developer. Many state DOTs focus on the state highway system in their databases. Consequently, they create nodes at intersections of state highways. However, the *TIGER* (Topologically Integrated Geographic Encoding and Referencing) databases have nodes where roads intersect local streets, political and administrative boundaries, railroads, utility lines, water features, canals, and so on. So, if you extract those arcs that are described as roads in the database, you get a lot of nodes at intersections of two arcs, and can clean these from the road layer (or coverage). You are left with arcs segmented at intersections with other roads (including local roads.)

Attribute Data Representation

In the early days of GIS-T, a basic problem arose because the segments (arcs) in the spatial database did not necessarily correspond to the segmentation scheme used to collect information about the transportation system. The segmentation scheme of arcs in the digital database depends on the scale of the source map. One issue faced by DOTs is that the scale of the digital centerline database is often too small (1:100,000 or 1:500,000) for easy correspondence with LRS. Obvious problems are missing facilities, such as ramps that do not appear on a 1:100,000 scale base map (Figure 2).

Within an agency, there may exist a variety of methods to segment the transportation system into discrete roadway sections on which to collect data (many of these methods are described previously). For instance, the pavement group may define control sections on the roads over which they collect a variety of condition characteristics, or as in many states roads are broken into fixed length sections (e.g. 0.1 miles) and all data characteristics are collected over each section. Each roadway section is associated with a single record in a tabular database and the linear references of the beginning and ending points of these sections are explicitly described unless the length of each section is the same. The characteristics of these fixed segments generally reflect average or typical values found within the section. However, this record may relate to several arcs or just a piece of a cartographic line segment in a geographic database.

Some 'inventory' databases are different from these 'condition' and 'performance' databases in that the roadway sections are not predefined. Each segment represents a unique combination of the roadway features (e.g. pavement condition, shoulder width, and average annual volumes). When any one of the characteristics changes, a new segment is created. In the literature, this approach is referred to as a variable length segmentation scheme. Again, each roadway segment is associated with a single record in a tabular database but under the assumption that all the characteristics stored in that record remain constant over that section. The linear references for the beginning and ending points of each segment must be explicitly stored in each record. Again, this record may relate to

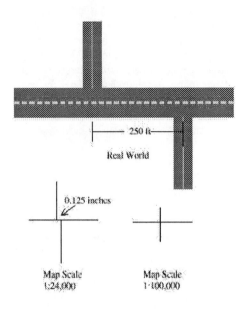

Figure 2   Feature representation.

several arcs or just a piece of a cartographic line segment in a geographic database. Both fixed (Figure 3) and variable (Figure 4) length segmentation approaches are criticized with regard to accuracy and data redundancy problems.

Integrating Spatial and Attribute Representations

GIS have been proposed as a suitable platform for implementing location reference systems. Therefore, GIS must be capable of storing and manipulating data based on a variety of linear location reference methods. The GIS-T community began looking for a method that could be implemented in the available GIS software which would allow for dynamically segmenting roadways according to the natural breakpoints of the attributes that one wanted to display and analyze. This was the beginning of the concept of dynamic segmentation (Dueker 1987, Nyerges and Dueker 1988, Nyerges 1990, Dueker and Vrana 1992).

To fully realize the power of dynamic segmentation, each roadway characteristic could be collected and maintained in a separate database according to the most appropriate and convenient segmentation. So, for example, the characteristic 'number of lanes' would be stored according to segments that begin and end only where there is an increase or decrease in lane numbers (similarly for pavement width, traffic volume, or degree of curvature data). However, it is not

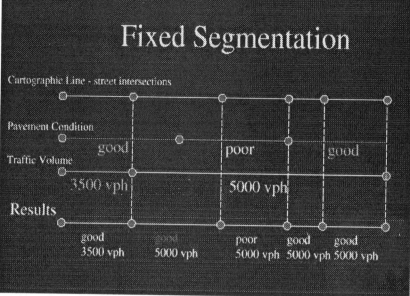

Figure 3 Fixed length segmentation schemes.

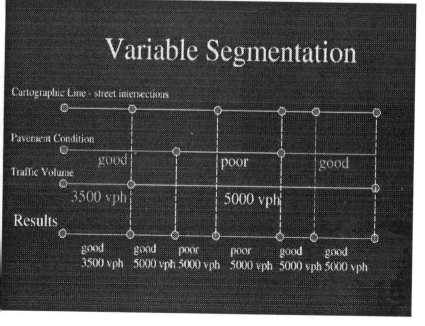

Figure 4 Variable length segmentation scheme.

convenient for DOTs to collect and maintain data in this manner. So, today many DOTs have a mixture of multi-attributed databases (inventory, pavement, safety, bridge, etc.) that use a non-optimal segmentation scheme to structure their data based on the fixed and variable length segmentation schemes. Dynamic segmentation capabilities within GIS and other software tools (such as dRoad by Deighton Associates) are used to integrate these databases. Today most GIS-T software has an implementation of dynamic segmentation. This segmentation relies heavily on the underlying linear referencing which has been implemented for the linear database. Each segment which is dynamically determined is identified with a route and a 'from' and 'to' distance that reflect the linear referencing method in place for the underlying network database.

Need for Linear Referencing in GIS-T

Linear referencing of transportation databases was in place long before GIS-T was developed. The referencing was used to determine locations of incidents and to manage and maintain the roadway facility. Implementation of LRS within a GIS-T requires that the route names (reference points, reference posts, or nodes and distance measures) be associated with the cartographic representation of the roadway.

The important capability of GIS-based linear referencing to share network data has been recognized by the federal geographic data committee (FGDC)'s ground transportation subcommittee (GTS). The GTS has recommended that a standard linear data model should be developed and that linear referencing be incorporated into the spatial data transfer standard (SDTS) transportation network profile (TNP). The following subsection gives an overview of the national spatial data infrastructure and the importance of linear referencing in sharing of transportation geo-spatial data. Coupled with GIS-T, linear referencing becomes a much more powerful tool for information management and information systems application development. The navigational and automated information needs of ITS implementations require a standard, easily communicated, and commonly understood linear referencing system. Next an overview of the need for linear referencing by the ITS community is given. As noted above, it is the linear referencing system that allows the dynamic segmentation capabilities to be implemented. Linear referencing and dynamic segmentation together provide the user with the ability to perform spatial analysis. It is also the capability that allows for the integration of data from different sources. The relationship between linear referencing and dynamic segmentation is discussed later.

SDTS Transportation Network Profile (TNP)

The national spatial data infrastructure (NSDI) was established by President Clinton in 1994 to promote the development and sharing of geospatial data among all levels of government, academia, and the private sector. The federal geographic data committee was charged with implementing the NSDI by coordinating the development of policies, standards, and partnerships to facilitate the production, sharing, and use of geospatial data nationwide. More information about the NSDI can be found on its website at http://www.fgdc.gov/NSDI/ Nsdi.html.

A key element in implementing the NSDI is the concept of framework data. Framework data consist of seven thematic layers most commonly used in creating maps and conducting spatial analyses. These thematic layers include: (1) geodetic control points, (2) digital orthoimagery, (3) elevation, (4) hydrography (surface water features), (5) governmental unit boundaries (e.g. states, counties, and urban areas), (6) cadastral (land ownership), and (7) *transportation*. Transportation features, particularly roads, are important not only for showing the relative density of development on a map, but also as a means by which other geospatial features can be located using address geocoding and linear referencing. In developing the NSDI, the FGDC is giving highest priority to creating and sharing databases representing these seven framework themes.

The FGDC's ground transportation subcommittee has the overall responsibility for coordinating the development of standards and procedures to facilitate the creation and sharing of transportation framework data. The GTS has undertaken several activities to develop standards, including drafts for a transportation network profile for the spatial data transfer standard and a transportation spatial data dictionary (TSDD) as a precursor to a more comprehensive transportation feature content standard. A position paper on the need for linear referencing as a required element in transportation framework databases has also been developed. These documents can be found on the subcommittee's website at http://www.bts. gov/gis/fgdc/.

As envisioned in the NSDI, framework data will consist of the 'best' available, non-proprietary geo-spatial data for a particular area, where 'best' is determined by a combination of geo-spatial accuracy, level of detail, completeness, and currency. Framework data will contain minimal attribution, leaving it to the end-user or value-added database resellers to attach attribute data needed for specific applications. However, framework data will include sufficient information to enable users to unambiguously identify specific transportation features for the purposes of attaching attributes and/or updating the spatial geometry.

The transportation theme presents some special problems with respect to developing and maintaining a framework data layer. First, transportation features, especially roads, are among the most dynamic of the framework themes. New roads are constantly being built, most of them in new developments in local communities. Also, many older primary and secondary roads are being realigned, resulting in changes to their length and shape. keeping a national road framework layer current will require contributions from numerous state and local transportation agencies responsible for overseeing road construction and acceptance.

Second, a transportation network is not a set of discrete, easily identified spatial features, but a more or less continuous feature that can be arbitrarily partitioned into an infinite number of segments, depending on the application. For example, cartographers may segment roads only at intersections, while state highway engineers may choose to partition them at changes in pavement or construction project limits, and local E-911 coordinators may want to identify each unique address as a network node. Without some means of being able to easily re-segment a network or attach application-specific attributes to it, the value of developing and sharing a single framework database for an area is lost.

Third, the long-term success of the NSDI depends on the ability to perform 'transactional updating' to the framework databases. Once users obtain a framework database for their area of interest and populate it with application-specific attributes, they are not going to want to receive an entire new version of the database as an update, particularly if only a few of the records have actually changed. Most would prefer to have the updated records identified and then be able to replace only those updated records in their existing database. To conduct transactional updating, each framework record needs some permanent identifier that remains stable over any version or subsequent re-segmentation. In order to address these problems, the GTS is looking to develop a standard linear data model that would enable users of a transportation framework database to re-segment and attach application-specific attributes regardless of how the framework database is segmented. Such a model should accommodate the various linear referencing methods currently in use by state DOTs and others, but should not inhibit local efforts from developing accurate and detailed cartographic databases that contribute to the NSDI transportation framework layer.

Intelligent Transportation Systems

The Intermodal Surface Transportation Efficiency Act (ISTEA) acknowledged that the nation had reached the goals set for the Interstate era and called for advancement into the next generation of surface transportation via a new level of research and exploration in pavements, design techniques, and intelligent transportation systems. The objective of ITS is the use of computer and communications technologies to improve safety, productivity, and general mobility in an era of increasing congestion, continuing threats to travel safety and security, and increasingly constrained transportation agency budgets. The definition of ITS used by USDOT is contained in the National ITS Program Plan, which identifies 29 specific user services - specific approaches to addressing specific transportation challenges - organized into six main program areas.

Those six main areas, which correspond to various offices and programs within USDOT, are responsible for delivering each component of the program. The areas are:

1. *Advanced Travel Management* - encompasses a wide range of ITS services that address traffic management, traveler information, transit management, and electronic fare/toll payment.
2. *Advanced Commercial Vehicle Systems* - apply advanced technologies and information networks to increase productivity and efficiency for both fleet operators and state motor carrier regulators.

3. *Advanced Crash Avoidance Systems* - comprise a program area that explores the use of ITS to improve driver safety through enhanced vehicle control, automated collision notification (ACN), and other technologies.
4. *Automated Highway System (AHS)* - comprises a program area that is defining system requirements for the fully automated highways of the future (a demonstration of feasibility has been presented in 1997 in San Diego).
5. *Rural Applications of ITS* - use innovative and advanced technologies to address numerous safety problems associated with rural transportation.
6. *System Architecture and Standards* - lay the foundation for national interoperability among all ITS components.

Each of these program areas implies a need for GIS and LRS, specifically the ability to quickly and effectively communicate information about the spatial location of events in terms which both people and computer systems can understand. Goodwin (1996) identifies all of the ITS data that need a location component transmitted. These include:

1. Traffic, incident data, and road advisories between state DOTs and Local Agencies
2. Traffic, incident data, and road advisories from state DOTs to regional information service providers
3. Warnings, status, and advisories between regional information service providers
4. Traffic and incident data between regional information service providers and local agencies
5. Inter-regional warnings, traffic and incident data between regional information service providers
6. Traffic and incident information, road advisories, guidance data, and tracking data between regional information service provides and vehicles
7. Database updates from the information service providers and map database vendors.

Because the need for GIS and LRS cuts across all these program areas, most of the work on GIS and LRS for ITS has occurred under the system architecture and standards program area. In this regard, efforts have been made to define an LRS that can be standardized for ITS use and which can serve the interoperability among all ITS components. The problem of developing a standard way of referencing location is compounded by the need to implement the method nationwide and at both national and local scales.

As part of the effort to define communication standards about location, a group of researchers have delved into LRS from the perspective of ITS needs. These include work done by Okunieff et al. (1995), Goodwin (1996), and Goodwin et al. (1995, 1997). Several LRM are evaluated along side the development of a standard location referencing message protocol (LRMP) for communicating the location reference. One final recommendation is to support multiple LRS for ITS while developing a National ITS Datum composed of an ITS Node Set and an ITS Geodetic Datum (Goodwin 1996). The key to the use of this system in GIS is the ability to interpret LRS as it is communicated by the LRMP and then to translate it to cartographic primitives.

Dynamic Segmentation

As noted above, it is linear referencing that allows one to implement dynamic segmentation in most GIS. Dynamic segmentation is described in detail by Dueker and Vrana (1992), and the need for it is clearly articulated by Hickman (1995). Hickman goes beyond the discussion of dynamic segmentation and suggests that a feature-based GIS will make dynamic segmentation unnecessary.

To associate events with sections of roadway or rail lines, most GIS models require that road segments (arcs) be grouped into routes. Each segment, that is a member of a route, shares the route's name (ID) and has information on the beginning and ending mile or kilometer point associated with the ends of the segment along the route. It is not necessary for an entire segment to be included in a route. Most dynamic segmentation models associate additional information with each segment in a route indicating what part of the segment belongs to a route. Routes do not need to be continuous. In other words, all segments in a route do not have to be connected to each other. Routes may not contain branches.

The route structures described above have been incorporated into many GIS software packages. However, transportation agencies have experienced several difficulties in manipulating these structures. The Wisconsin DOT has developed their own data model and algorithms to manipulate routes and segments (arcs). These modifications were necessary to support their application needs. The Wisconsin model is documented in the LRS in GIS CD product developed by the Bureau of Transportation Statistics.

CONCLUSIONS

If linear referencing methods are standardized across agencies and applications, then it becomes possible to integrate data using dynamic segmentation from disparate sources. For example, the locations of pavement condition ratings and counted volumes can be overlaid on the same linear referencing system to analyze the places where the worst pavement conditions are experienced by the most number of vehicles. As another example, if the travel demand model networks of an MPO are linearly referenced with a standardized system, then forecasted volumes can be combined with state-level inventory data to calibrate the model based on HPMS counted volumes. All these examples require that a linear referencing system be in place and that it be standardized across organizations and applications. This linear referencing system must support multiple linear referencing methods. The NCHRP project 20-27 research results provide a data model that supports multiple linear referencing methods associated with multiple cartographic databases (NCHRP 1997). The Wisconsin DOT has made great strides toward implementing this model.

These topics and others are covered in the LRS in GIS CD product developed by the Bureau of Transportation Statistics. The CD contains over a hundred original documents detailing how state DOTs are implementing LRS in GIS as well as research papers on theory and applications of LRS in GIS. For more detailed treatment of this topic, the LRS in GIS CD may be downloaded for the BTS website at www.bts.gov.

REFERENCES

Deighton, R. and Blake, D. (1994). "Improvements to Utah's location referencing system to allow data integration." Transportation Research Board, *Third International Conference on Managing Pavements*, 1, 97-107.

Dueker, K.J. and Vrana, R. (1992). "Dynamic segmentation revisited: a milepoint linear data model." *GIS-T Symposium Proceedings. American Association of State Highway and Transportation Officials*, 63-78. (Also, Journal of the Urban and Regional Information Systems Association, Volume 4, Number 2.)

Goodwin, C.H. (1996). "Location referencing for ITS." *Materials for the Linear Referencing and the Spatial Data Transfer Standard Workshop*, Sponsored by the Bureau of Transportation Statistics, Washington, DC.

Goodwin, C., Gordon, S., and Siegel, D. (1995). "Reinterpreting the location reference problem: A protocol approach." *GIS-T Symposium Proceedings, American Association of State Highway and Transportation Officials*, 76-88.

Goodwin, C., Gordon, S., and Siegel, D. (1997). "Deploying the location reference message specification." *ITS America Seventh Annual Conference Proceedings*.

Hickman, C. (1995). "Feature based data models and linear referencing systems: Aids to avoiding excessive segmentation of network links." *GIS-T Symposium Proceedings, American Association of State Highway and Transportation Officials*, 89-113.

National Cooperative Highway Research Program. (1974). *Highway location reference methods*. Synthesis of Highway Practice Number 21. TRB, National Academy of Sciences, Washington, D.C.

National Cooperative Highway Research Program. (1997). *A generic data model for linear referencing systems*. NCHRP Project 20-27 (2). NCHRP Research Results Digest Number 218, TRB, National Research Council, Washington, D.C.

Nyerges, T.L. (1990). "Locational referencing and highway segmentation in a geographic information system." *ITE Journal*, 27-31.

Nyerges, T.L. and Dueker, K.J. (1988). *Geographic information systems in transportation*. Office of Planning HPN-22. Federal Highway

Administration, U.S. Department of Transportation, Washington D.C. (Reprinted in Regional Computer Assisted Cartography Conferences: Summary, August 1988).

Okunieff, P., Siegel, D., Miao, Q., and Gordon, S. (1995). "Location referencing methods for intelligent transportation systems (ITS) user services: Recommended Approach." *GIS-T Symposium Proceedings, American Association of State Highway and Transportation Officials*, 57-75.

Star, J. and Estes, J. (1990). *Geographical information systems: an introduction.* Prentice Hall, Englewood Cliffs, N.J.

Worboys, M.F. (1995). *GIS: a computing perspective.* Taylor and Francis Ltd., London, United Kingdom.

Section 2

GIS Applications

CHAPTER 6

Regional Planning
(Activity-Allocation Modeling)

Larry G. Mugler and Terence T. Quinn

A fundamental activity of regional planning programs is the development of long-range forecasts of activity. These may relate to wastewater flows, transportation levels on major roadways or transit lines, the number of elderly needing public services, or a number of other issues. Each of these issue areas depends on the forecasts of demographic activity to predict needs in its area. This chapter discusses the way one regional planning agency, the Denver Regional Council of Governments (DRCOG), has utilized its geographic information system (GIS) to assist in the preparation of demographic forecasts.

OVERVIEW

Future household and employment levels for small geographic areas can be predicted in a variety of ways (Haring et al. 1992, Verburg et al. 1997). Straight-line extrapolation of past trends can be improved by more sophisticated mathematical curve fitting routines. Multiple regression analysis can be used to relate other variables to population and employment growth. Some regional agencies use elaborate group processes to involve local governments in determining the likely pattern of future growth. Several sophisticated simulation models have been developed, such as DRAM/EMPAL (Putman 1983) and MEPLAN (SPARTACUS 1999). The Denver Regional Council of Governments (DRCOG) model is an attempt to use some trend information, but to add policy variables to offset the assumption that the future will be a repetition of the past. For other examples of regional activity models, see Bell et al. (1997), Klosterman (1999), Seskin (1994), and Wegner (1994).

Several important factors were addressed in developing the regional activity allocation model (RAAM). First, a method was needed that could adequately model alternative urban development patterns that are significantly different from historical trends. Second, the method needed to be systematically and consistently applied across the study region. Third, the method needed a sound theoretical basis for determining the relationships within the model. Finally, the model had to be easily updated between iterations to reflect different policy options.

The process utilizes both top-down and bottom-up approaches to combine professional knowledge and field information in an attempt to predict the region's future. As an allocation model, external control totals are generated through an econometric model for the region as a whole and distributed among a set of subregions (traffic analysis zones). The information about the zones is derived from existing sources such as the U.S. Census and from local knowledge about the characteristics of the zone. The major steps include the development of subregional control totals, the creation of a zonal attractiveness index (ZAI) for comparing zones with each other, and the distribution of the control totals among the zones based on their relative attractiveness.

STRUCTURE OF DRCOG MODEL

As noted, the RAAM model starts with regional forecasts of population and households. These are prepared from national econometric models and relate the region's economy to that of the nation. After forecasting the growth of the economy, in particular the jobs by industry, the model estimates the labor force needed to fill those jobs. A standard population cohort model is used to determine the migration levels needed to produce the labor force over the forecast period, in our case from 1990 to 2020. Census Bureau models are used to predict the number of households based on the forecasted population by age groups.

The Denver region has been divided into five subregions that have relatively unique characteristics: the central business district, the inner ring of historical development, the growing suburban fringe, several free standing communities (such as Boulder), and the rural portions of the region. A control total for jobs and households is established for each of these subregions based on the policy direction of the regional plan. For example, the currently adopted forecast assumes that the central business district will grow from 100,000 to 150,000 jobs in thirty years and add 20,000 households to an existing base of about 5,000.

The region has been divided into 1503 traffic analysis zones as the most detailed unit for forecasting and analysis. The subregions contain from 36 zones to over 400 in the fringe area. The RAAM model needs to know several characteristics about each zone to carry out two tasks. First, the model develops a zonal attractiveness index for each zone. The index is made up of seventeen variables, many of which are generated using GIS techniques and are described later in this chapter. These variables can be weighted and evaluated by a number of 'if-then' statements to rank order the traffic analysis zones (TAZ) as to their suitability for development. By changing the weightings and the 'if-then' statements for various development scenarios, the process allows for consistent, predictable, and relatively rapid development of alternative datasets.

Second, once the ZAI is calculated, the characteristics of the zone are used to fill the zone with households and/or jobs. By knowing the mix of existing uses, their densities, the amount of vacant developable land and the planned uses (such as transit stations) by TAZ, the model can systematically allocate households and employment to each zone based on a variety of strategies. The key to the model is the assumption that only the most attractive zones in each subregion will be totally filled. The remaining zones are filled only to the extent that their ZAI relates to the ZAI of the most attractive zone. For example, a zone with a ZAI of 500, when the most attractive zone has a ZAI of 1,000 would only be allowed to fill one half of its vacant land.

DATA PREPARATION AND ANALYSIS

Using GIS in Defining Modeling Areas

The focus of this chapter is not on the model itself but on the variety of ways in which GIS has made the development of the model possible. Without GIS, the analysis of 1,500 TAZ would be entirely too burdensome. Past models were based on simple regression models for 50 subregions that were then disaggregated to zones manually. By using GIS, the TAZ could be tested for a variety of characteristics and the RAAM spreadsheet model developed. This section looks at several of the GIS techniques used to prepare the TAZ for modeling.

Creation of Zone Polygon Coverage

In order to use GIS methodologies, the TAZ needed to be re-created within the GIS system. For DRCOG, this effort began many years ago when GIS programs were less sophisticated. Our original purchase of GIS software (the ATLAS system) came with geographic coverage for 1980 Census tracts. Subdividing the tracts into zones created our initial TAZ coverage. Zone and tract boundaries were drafted onto 1:24,000 U.S.G.S. quadrangle maps that were mounted on a digitizing tablet. Zones in almost all cases nest within tracts so that only the boundary feature such as a road or stream needed to be directly digitized from the map. This line feature could then be used to split the tract polygon into the zone polygons. This ability to use GIS software to create new polygons without digitizing every boundary element saved significant amounts of staff time and money.

Even as zones have been revised over the years, this initial coverage has formed the basis for each new generation. Following the 1990 census, the current set of 1,503 zones was defined and the GIS coverage updated. It should be noted that this approach results in a zone coverage that is only as accurate as the original tract coverage. We are now in the process of obtaining new road coverage with major improvements in positional accuracy. Nodes in the new file are expected to be accurate within 5 m while the old file may be as much as 30 m off. Updating the zone coverage is expected to be a major work task over the next year.

Demographic Attributes

Having the zones created in a GIS layer or coverage provides only limited data such as area and perimeter. To be useful, the GIS system needs to provide attributes about the zones that represent steps in forecasting future TAZ activity. Fortunately GIS software is designed to efficiently create attribute files. The example described here is the development of demographic information for the TAZ. For a discussion of basic demographic theory, see Yaukey (1985).

One of the more powerful GIS functions is the ability to aggregate data from smaller geographic units into larger units. Since the major source for demographic data is the U.S. Census, the TIGER (topologically integrated geographic encoding and referencing) files were brought into the DRCOG GIS system as a foundation for demographic analysis. The basic unit for census demography is the block, which has attributes such as number of households, number of persons and ethnicity. The TAZ coverage was placed on top of the block boundary coverage and two different approaches to summarizing data were applied. For another example of this application of GIS, see Sathisan et al. (1998).

First, the GIS system was used to build a block to zone equivalency file. The system took each zone and determined which blocks had a centroid that fell within the zone. It then created a table showing the blocks per zone. We were then able to use this equivalency file in a database to aggregate any data collected by block. This was especially important for post-census data such as current employment estimates by place of work. This information was originally geocoded to census block that allowed us to match the two databases and sum the employment by zone.

The second approach used the power of the GIS system more directly. The 1990 Census file by block was loaded into the GIS program as attribute coverage. The GIS program not only could provide the centroids of the zones within a tract, but could also sum the demographic characteristics of the blocks into a zone attribute table. This bypassed the database functions entirely.

Land Use Attributes

To forecast future activity within a zone, it is necessary to know some basic land use attributes about the zone. For example, does the zone have any vacant and developable land? Is the developed land within the zone mostly residential, mostly commercial, or a mix of uses? Are there significant open space or other constraints to development? The GIS system can be used to develop a land use attribute table for the TAZ.

Since DRCOG's land use information for this process had not been created in any digital form, we have not been able to fully utilize the GIS system to define the current land use in the TAZ. We were able to use the tract to TAZ equivalency to take an older land use table by tract and disaggregate it to TAZ and this data was reviewed by local planning departments to locate major errors in the allocation process. For the future, we have purchased digital orthophotographs of the region, which we can bring into the GIS system. We can then lay the TAZ boundaries directly on the photos on a monitor and digitize land uses (such as the amount of vacant land) without ever using a drafting table or digitizing tablet.

Development of Zonal Indices

DRCOG staff was able to utilize GIS technology to derive many of the indicators used for the RAAM. GIS operations allowed for speedy spatial data calculations, drastically reduced the resources required to produce the indices and allowed for the development of innovative growth indicators. In short, GIS technology presented a new and powerful tool by which to approach policy and scenario modeling.

DRCOG staff began developing model indices by entering into a collaborative effort with planners from many of the regional jurisdictions. The purpose of this work was to determine what TAZ attributes would most likely attract new household and employment growth. DRCOG equated the agreed upon attributes to measurable factors that became the TAZ indices. Since DRCOG has been using GIS for sometime, it had much of the spatial and statistical data required for index creation. Yet, staff still found it necessary to collect new information and convert it to usable GIS data. Many GIS operations were used alone, and in series, to help weigh the effects that each index potentially has on future growth. This section of the chapter highlights the different types of GIS activities used in the development of the zonal attractiveness indices.

Point-in-Polygon Operation

The point-in-polygon operation was used in series with other GIS operations and statistical procedures to aid in the development of current household and employment estimates. These estimates are used to develop projections for the year 2000. The RAAM has an Employment and Population Index that measures the rate of change between the year 2000 projections and 1990 Census data. The year 2000 is the jumping off point for the model that allows already planned development to be completed, and thus influence future growth, before the RAAM begins the allocation process.

GIS processing is used to create current population and employment estimates from State income tax roles, Census data, electric utility records, and privately developed databases. These data are geocoded on the TAZ polygon layer and the road base map. The point-in-polygon operation is used to sum the employment and population by TAZ. The TAZ polygon layer is laid over the geocoded employment and population data. The point-in-polygon operation is then used to sum TAZ totals. TAZ employment data is summed by business addresses that are geocoded to a road layer. TAZ population totals are summed through the aggregation of Census block geography centroids (Figure 1). Planners and developers then meet to discuss what completed projects, population totals, and employment totals will be added to the current data to achieve 2000 data. The rate of change for employment and population between 1990 totals and the new 2000 totals are figured, broken into 10 percent intervals, and are expressed as a TAZ index score on a scale from 1 to 10.

Polygon Overlay Operation

The polygon overlay operation was used to measure how much of a TAZ is influenced or affected by particular indicators such as environmental constraints and the extent of infrastructure. The environmental constraints index simply asks how much of a TAZ's development potential is constrained by environmental attributes. Environmental attributes include natural resources like coal deposits, natural hazards such as flood plains, and hazardous materials such as nuclear

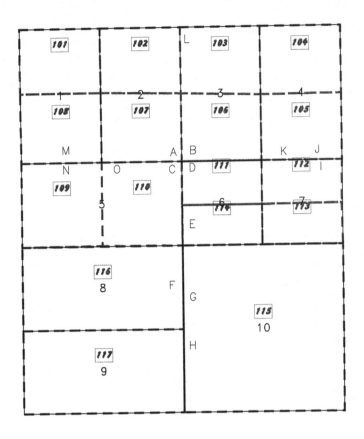

Figure 1 Point-in-Polygon

weapon manufacturing facilities. Data on subjects such as these were collected and digitized to form the environmental constraints layer for the entire Metro Area. The constraints polygon layer was then electronically laid over the TAZ layer. The overlay operation was initiated to determine how much of each TAZ is covered or influenced by the environmental constraints layer. The percentage of the TAZ that is covered is subtracted from 100. This new percentage demonstrates which TAZ are least affected by the environmental constraint indicator and are therefore more attractive to growth.

The infrastructure index was developed in much the same way as the environmental constraints index. Mapping and combining all of the existing wastewater service provider districts in the Metro Area achieve a polygon layer of served areas. It is assumed that current wastewater services are an excellent proxy for the extent of urban service infrastructure. The infrastructure polygon layer was laid over the TAZ layer and then the overlay operation was initiated (Figure 2). The results of the operation produce the area of each TAZ that falls within the boundaries of the wastewater service area. This is turned into an index by determining the percentage of each TAZ's area that is served by urban infrastructure. The more infrastructure coverage a TAZ has, the more growth it potentially attracts.

Buffer and Area Inside Operations

Many of the indices developed for the RAAM utilized the buffer and area inside GIS operations. The buffer operation was used to create areas of impact or influence around GIS map points, line segments, and polygons. Once the buffer areas of influence are created, the TAZ layer is turned on. The overlay operation is initiated to measure the percentage of the influence area that lies beneath each TAZ. The buffer and area inside operations used in series is an excellent way to measure how much influence urban attributes like light rail stations (LRT), regionally significant roads, and high density urban centers have on potential areas of future growth.

Indices were created by using the buffer operation around various point, line, and polygon layers. The station index utilized a point layer that represented proposed regional LRT stations (Figure 3). This index expresses the desirability of development to locate near LRT stations because they are planned to have high levels of activity and easy access to services. Therefore, each TAZ that has some percentage of its area covered by a station influence buffer would be more prone to attract growth. A half-mile radius buffer expresses the area of increased activity around the LRT station points. The overlay operation was initiated to measure the area of each TAZ underneath the station activity areas as a percentage of the total TAZ. A large percentage of station activity area per TAZ reflects increased levels of future growth. The TAZ operation results were grouped by 10 percent intervals and expressed as an index score on a scale from 1 to 10.

The roads index was developed to demonstrate the impact major roadways have on the attractiveness of a TAZ (Figure 4). These popular roads attract businesses to them because of high levels of activity. Residential uses try to locate nearby because of access and convenience. Regionally significant roads are defined as the major collectors, arterials, and highways that are an integral part of the regional transportation network. The index applies a half-mile buffer to the line segments that represent regionally significant roads. The TAZ area that lies underneath the road influence buffers increases the TAZ's chance of attracting future growth. The area inside operation is used to measure the percentage of a zone that lies beneath the roads influence area. The result is grouped into 10 percent intervals and is expressed as a score for each zone on a scale from 1 to 10.

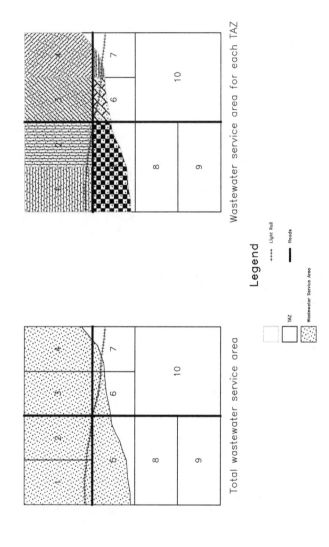

Figure 2 Overlay (wastewater sevrice area for each TAZ)

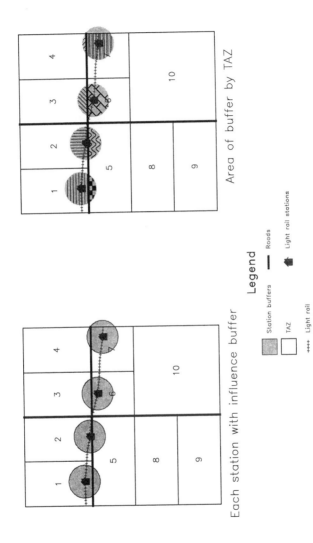

Figure 3 Buffer and area inside operation series (point)

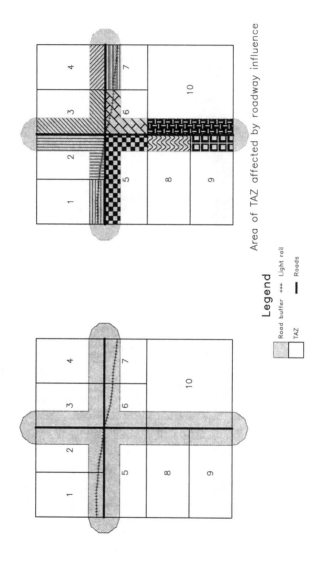

Area of TAZ affected by roadway influence

Legend

Road buffer +++ Light rail

TAZ

Roads

Figure 4 Buffer and area inside operation series (lines)

The urban center index represents a series of TAZ that are considered to be the locations of future high intensity development. Influence areas were created for each of the center TAZ because of the intensive development that would spread or spill over into neighboring TAZ. Urban center zones are modeled as being attractive for higher density employment and population growth. The area around these centers would experience similar development, just at a lesser scale. The buffer operation was initiated to measure the spill over effect that would take place within a half-mile of the designated urban center TAZ (Figure 5). The area inside operation is then used to measure the percent of a TAZ area that is covered by the urban center influence buffer. The results are grouped into 10 percent intervals and are expressed as a score for each TAZ on a scale from 1 to 10.

Nearest Neighbor Operation

The nearest neighbor operation provided the opportunity to represent zonal attractiveness based upon its proximity to particular features. The open space index was developed to demonstrate that a TAZ is more likely to attract future growth based on how close it is to open spaces (Figure 6). The DRCOG open space system was digitized based on the commitments by member jurisdictions to maintain various natural areas, resource areas, trail systems, and parks as open space. The open space polygons were added to the regional zone layer. The distance operation was used to determine the distance from each TAZ's centroid to the closest open space polygon centroid. The longer the distance, the less desirable the zone is for development. The ordinal distances for each TAZ were grouped into 10 nominal quartiles, which are used as the index scores ranging from 1 to 10.

SUMMARY OF RESULTS

The previous sections have highlighted several GIS technologies used to develop indices to determine the attractiveness of TAZ to development. Table 1 summarizes the data from the sample indices to show how an overall ZAI might be determined for the 10 sample zones. In this example, all seven of the indices are given equal weight. In the actual DRCOG model, a committee of local and regional planners and economists weighted each of the 17 variables, as they believed the index relatively influenced population or employment growth. From this exercise, zone 3 is the most attractive and zone 9 is the least.

Another GIS methodology should be noted here - thematic mapping. Figure 7 uses the ZAI scores to illustrate the locations of the high and low attractive zones. The planning staff uses figures such as this to evaluate the overall success of the ZAI scores in measuring attractiveness. Looking at this figure may suggest that the rail index is under weighted or that open space is over weighted and the staff can adjust the formula to reflect their understanding of the development process.

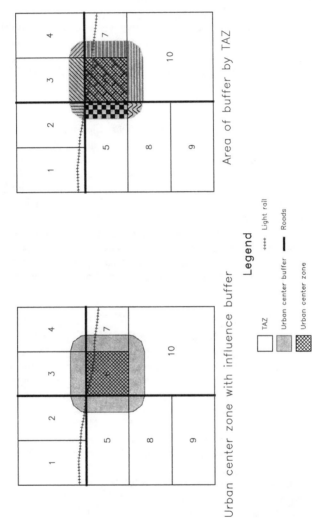

Figure 5 Buffer and area inside operation series (polygon)

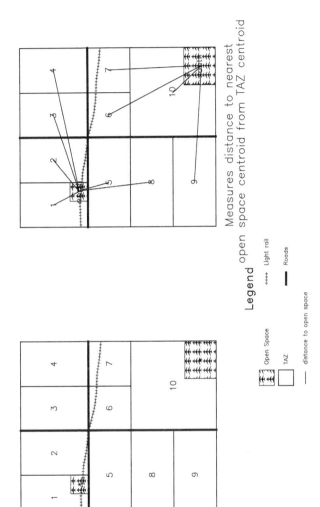

Figure 6 Nearest neighbor operation

Table 1 Summary of zonal attractiveness indices

TAZ	Corresponding Census Blocks	Census Population	Population		Corresponding Employment Locations	Employment			Wastewater	
			Total	Rank		by Location	Total	Rank	Service	Rank
1	101,108	80+50	130	9	M	100	100	9	100%	10
2	102,107	30+20	50	4	A	70	70	7	100%	10
3	103,106	90+90	180	10	B,L	35+90	125	10	100%	10
4	104,105	30+25	55	5	J,K	10+20	30	5	97%	7
5	109,110	40+50	90	8	N,O,L	10+10+15	35	6	73%	6
6	111,114	20+10	30	2	D,E	15+10	25	4	35%	5
7	112,113	30+40	70	7	I	5	5	2	14%	4
8	116	70	70	7	F	15	15	3	0	3
9	117	30	30	2	-			1	0	3
10	115	40	40	3	G,H	25+50	75	8	0	3

TAZ	Station Buffer		Roads Buffer		Urban Center		Nearest Open Space		Total	TAZ Order
	Area	Rank	Area	Rank	Area	Rank	Distance	Rank	Rank	Of Attractiveness
1	21%	8	23%	5	0%	2	0.73	10	53	2
2	15%	7	51%	9	6%	5	1.01	7	49	3
3	7%	5	51%	9	23%	8	1.98	5	57	1
4	2%	4	23%	5	6%	5	3.04	1	32	8
5	13%	6	48%	7	18%	7	0.77	9	49	3
6	31%	9	60%	10	100%	10	2.66	3	43	5
7	34%	10	36%	6	36%	9	2.38	4	42	6
8	0	3	18%	3	5%	3	1.85	6	28	9
9	0	3	18%	3	0%	2	2.83	2	16	10
10	0	3	18%	3	12%	6	0.99	8	34	7

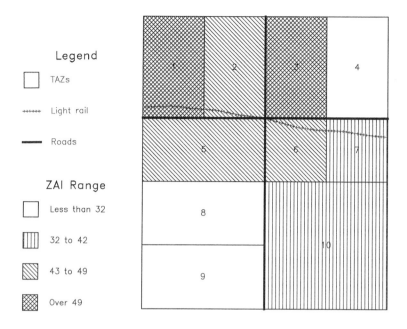

Figure 7 Summary of zonal attractiveness scores

CONCLUSIONS

GIS techniques are valuable tools for regional planners as they attempt to determine the future growth of a metropolitan area. They can be used to aggregate or disaggregate existing data. They can overlay one type of data on another. They can be used to display data for presentation and analysis. While DRCOG has been using GIS for many years, the tools are continually improving. The example of the digital orthophotographs is only one example of the ways in which GIS systems will continue to improve the forecasting and analysis process.

REFERENCES

Bell, M., Nicolson, K., Blake, M., and McQueen, I. (1997). "Forecasting population and housing development for small areas using GIS." *23rd General Population Conference*, Beijing, China.

Haring, L.L., Lounsbury, J.F., and Frazier, J.W. (1992). *Introduction to scientific geographic research*. McGraw-Hill: New York, N.Y.

Klosterman, R.E., Brail, R.K., and Bossard, E.G. (eds.). (1993). *Spreadsheet models for urban and regional analysis*. Center for Urban Policy Research: New Brunswick, N.J.

Klosterman, R.E.(1999). "The what if? collaborative support system." *Environment and Planning B: Planning and Design*, 26 (in press).

National Center for Geographic Information and Analysis. (1999). Internet: http://ncgia.ncgia.ucsb.edu:80

Putman, S.H. (1983). *Integrated urban models: policy analysis of transportation and land use*. Pion, London, United Kingdom.

Sathisan, S.K., and Rasas, J.P. (1998). "On the allocation of census data to traffic analysis zones for travel demand forecasting." In: Said Easa and Donald Samdahl (eds.). *Transportation, Land Use and Air Quality: Making the Connection*. Proceedings of a specialty conference, sponsored by American Society of Civil Engineers, Portland, Oregon.

Seskin, S. (1994). "The LUTRAQ project: travel behavior." *Presented at the Lincoln Institute of Land Policy Conference*, Portland, Oregon.

SPARTACUS (1999). "System for planning and research in towns and cities for urban sustainability." Internet: www.ltcon.fi/spartacus/default.htm

Verburg, P.H. and Veldkamp, A. (1997). "Modeling the spatial pattern of land use change in China." In: P.W.J. Uithol and J.J.R. Groot (eds.). *Proceedings Workshop Wageningen-China*, Report 84 AB-DLO, Wageningen, The Netherlands.

Wegener, M. (1994). "Operational urban models: state of the art." *Journal of the American Planning Association*, 60(1), 17-30.

Yaukey, D. (1985). *Demography: the study of human populations*. St. Martin's Press, New York, N.Y.

CHAPTER 7

Transportation

Reginald R. Souleyrette and Tim R. Strauss

The use of geographic information systems (GIS) in the transportation sector, commonly referred to as GIS-T, is explored in this chapter, focusing on those GIS applications found in urban planning and development. After a short section on applicable technologies and methodologies, two general types of applications are discussed: transportation planning and traffic engineering. Not all applications fall cleanly into one of these categories (e.g. traffic impact analysis, intelligent transportation systems), but the classification serves to focus the following discussion on the multitude of potential GIS applications in the transportation sector.

INTRODUCTION

This chapter attempts to provide the reader with an idea of how GIS technology is being used to solve problems in urban transportation. The focus is on applications within transportation planning (e.g. transit planning, travel demand management, network models, and freight transportation planning) and transportation engineering (e.g. traffic control and inventory, safety analysis, and intelligent transportation systems). Although many useful applications in the private transportation sector are relevant to urban planning and design, they are not covered here. These include applications typically found in the intercity air, rail, and trucking industries (e.g. routing, scheduling, tracking, location-allocation, and fleet and logistics management applications).

The potential for research and application in the transportation area is largely untapped but is likely to grow (DeWitt and Ralston 1996, ESRI 1999). GIS applications in infrastructure management and linear referencing systems integration with GIS are covered in Chapters 5, 7, and 8. In this chapter, we focus on GIS applications in transportation planning and traffic engineering. However, we do not identify all possible GIS applications in these areas, especially as many of these applications are continually being developed. The interested reader is encouraged to investigate the resources referenced throughout the chapter.

APPLICABLE TECHNOLOGIES AND METHODOLOGIES

A variety of agencies use or could use GIS for transportation applications. Typically involved in urban transportation are public works departments, local and state traffic engineering departments, local and sometimes state transit authorities, metropolitan or regional planning organizations (MPOs), councils of governments (COGs), state departments of transportation (DOTs), consultants (planning, traffic engineering, or transit), and universities. Despite their diversity, many applications of these agencies are related and can or do make use of common tools and databases.

Transportation departments and related agencies use GIS to undertake a variety of activities. Some of the functions that GIS addresses within these agencies (such as database integration, data query, and data management), and some of the specific techniques it performs (such as buffers and overlays) are common to other types of agencies as well. However, certain GIS activities are more specific to the concerns of transportation. In particular, network analysis is of key interest to transportation-related agencies. Its development within GIS is relatively new compared to the development of other forms of spatial analysis within GIS, but recently it has gained much more attention.

GIS for network analysis within a transportation agency can take several forms, many of which can now be performed on relatively simple software packages. Shortest path analysis can be used to find the route that minimizes the distance (or time or cost) between two points on a network defined to comply with certain criteria, such as lane width or the absence of highway-rail at-grade crossings. This basic form of network analysis capability can be used in several ways, such as to support permit routing for oversize/overweight vehicles, or to assist emergency vehicle response. This type of analysis can be extended, in shortest delivery route analysis, to incorporate several intermediate stops. These stops can be visited in a specific order and with specific delivery times. This functionality can be used by local transportation firms to plan delivery schedules and by transit agencies to develop bus routes or demand responsive paratransit routes and services.

Network analysis can also be used to find the nearest facility for a given purpose and the best route to follow. This technique can be used, for instance, to allocate students to the nearest elementary school or highway crash victims to the nearest hospital. Conversely, travel time or 'catchment' areas around a facility can be used to find, for example, all areas within 10 minutes of a fire station, airport, or school. The analysis of catchment areas, and the characteristics of the populations within these areas, is very useful in assessing service regions and potential capacity problems. The surrounding population can be allocated to the most appropriate facility, subject to travel time and capacity constraints.

GIS is generally not designed for full-scale network analysis in the form of urban transportation models. Several software packages, such as Tranplan and QRS II, are available for that specific purpose. The integration of these packages with GIS can occur through the exchange of network data and the calling of analytical routines across software packages.

Network analyses require high quality data, and a good deal of preparation is often necessary before the analysis can be conducted. Extensive and accurate information is needed on the connectivity of the network, turning options at intersections, network capacities, and demands on the system. The implications of poor quality data, such as missing or incorrect line segments, incomplete node information, and inaccurate capacity and demand data, are perhaps more serious in transportation network analysis than in other types of GIS activities.

TRANSPORTATION PLANNING APPLICATIONS

The late 1980s saw the first widespread use of GIS in transportation, mainly in the planning community. Engineers had been using CAD (computer aided drafting or design) tools for some time, but the early applications of GIS were implemented in scales and accuracy below the standards of most engineering applications. Therefore, a debate ensued (and continues today, to some degree) between planners who wanted and could use rapid development of GIS at relatively small scales (1:100,000, or 1:24,000) and engineers who required larger scale mapping on which to base their designs (1:1200 or better).

At issue is whether to proceed with inferior base maps and reap the benefits of early GIS deployment or to defer these benefits until high quality base mapping becomes available, thereby eliminating redundant data collection and processing efforts. Moreover, as GIS differs from CAD in that it can create new data, early decisions about scale and accuracy can have long-lasting effects. As USGS and Census maps were available for many areas, the first applications of GIS were naturally for planning purposes. With the recent advent of high quality, digital aerial ortho-photography, improved remote sensing, and GIS systems built on CAD engines, GIS can now be found in the engineer's toolbox as well. In any case, the goal should be to create a common network database for transportation, traffic engineers, and public works analysts for cross-disciplinary decision support, the ultimate aim of any GIS-T.

Even if decision support is the simple goal of a transportation planning GIS, several models and datasets are available in the planner's kit of tools. Modeling is largely a mature science in transportation planning. Planners model trip generation for impact studies, land use projection, population and employment projection, four-step traffic projection for network traffic assignment and corridor studies, air quality assessments of mobile sources of pollution, alternatives analyses, noise analysis, and energy studies, among other applications (Denno and Brail 1993). Databases to support these models include hourly and daily traffic volumes, geometric and signal data, travel survey data, existing and historical land use information, land development indicators such as building permits, and historical and current population/demographic data.

While GIS can be used for these applications individually, the integration of these models and datasets in a framework to effectively support decision making is a greater challenge, particularly as many decision makers are unaccustomed to the terminology and limitations of the approaches. However, recent developments in GIS (object-oriented programming and web-based customization tools, full-motion video, etc.) should be helpful in the design of such a framework. The following sections describe some of the applications of GIS to transportation planning that can be found in the literature.

Sketch Planning and General Use

A recent trend in GIS-T is the integration of analytical procedures into GIS by an end user or consultant. For a given analysis, four scenarios are possible:

1. The analytical functionality is included, off the shelf, without the need to purchase additional software modules,
2. The analytical functionality is available, but the user must purchase an additional module or package
3. The analytical functionality must be added to the GIS by customizing through its macro development language (if it exists)
4. The analytical functionality exists in another package outside the GIS and data must be imported between the two systems.

Along the lines of scenario 3, Bailey and Lewis (1992) present a methodology for creating a municipal GIS-T for a small to medium-sized area. They report the modification of a GIS to provide traffic assignment, vehicle routing, reapportionment of transportation analysis zones (TAZ), and location of facilities on a network. They state that "Building simple transportation tools directly within a general GIS package may be an appropriate avenue for smaller agencies without sufficient staff and resources to support multipackage solutions." The key to the success of these projects is the availability of data sources, as applications that require a lot of care and feeding will meet with uncertain success.

Baran et al. (1993) describe the implementation of GIS in the Detroit MPO. An implementation plan is presented that includes development of GIS layers based on TIGER/Line files. TIGER Tract/Block boundary files were used to develop TAZ and other thematic maps. The second phase of the project focused on the enhancement of the TIGER files for geocoding, which resulted in the development of site maps, travel survey maps, economic activity maps, and other site specific maps. A third phase saw the completion of a conflation exercise to integrate travel demand model (Tranplan) and crash data records. This resulted in the creation of accident, economic activity, demographic, and linear vehicle miles traveled (VMT) maps. The final phase of the effort investigated the use of GPS and aerial photo technology to improve the positional accuracy of the TIGER base map. It is important to note that the process of conflating data from one spatial database to improved cartography reduces or eliminates duplicating effort after higher quality base maps become available.

Murakami (1997) showed that GPS could be used with handheld computers for household travel surveys. User interfaces were designed to limit obstacles related to the use of new technology. Survey participants preferred using the computer rather than written logs by almost a nine-to-one margin.

Transit Planning and Operations

Transit is one of the more active areas for GIS implementation in transportation planning (Allen et al. 1993). Studies have specifically identified GIS component tools that are useful in transit applications. Antonisse (1991) lists geocoding, overlay, buffering, and route data structures as key features. These foster efficient determination of population characteristics within route service areas, route maps consistent with underlying street geography, most efficient transit routes, minimum travel cost itineraries for maintenance crews, and demographic trends affecting introduction of new services. It is noteworthy that some transit providers began using in-house GIS systems based on GBF/DIME (predecessor to TIGER) files as early as 1980. Many of these proprietary systems have since been replaced by off-the-shelf or customized applications.

Transit Demand

Recognizing the potential for GIS in visualizing transit demand, Azar and Ferreira (1992) discuss integrating GIS and transit modal split and demand models. In this effort, GIS was used to identify clusters of city blocks that house families with socioeconomic and demographic characteristics conducive to transit ridership. Benefits of visualizing routes and demographics/employment data include better targeting of high transit ridership areas in the design of route alignments.

Marketing

GIS has also been used in the marketing of light rail transit in Baltimore (Sirota and Henry 1995). Data from a passenger survey, census data, and local demographic forecasts were imported into a GIS for spatial analysis and effective color displays for senior management. In this effort, GIS was used to identify TAZ with the highest potential for new ridership.

System Management

The Montreal transit authority developed a special GIS to process origins and destinations based on client specifications. Multiple forms of spatial referencing were provided, including street addresses, trip generators and attractors, monuments, special activities, street intersections, and transit references (e.g. subway and rail stations, bus routes, and terminals). The best path calculation is carried out using interactively calibrated impedance functions (walking, in-vehicle, waiting, transfers, fares, and mode restriction). The authors report that an added challenge was the necessary consideration of the network geometry, commercial speed, day type, trip time and headway, and the validation of the computed paths using existing schedules. They also report that for a transit network (the size of Montreal's) consisting of 16,000 bus stops, 230 bus routes, and a pedestrian network of 80,000 links and 28,000 nodes, the calculation of the optimal path for a given origin-destination (O-D) pair is less than 2 seconds using a typical PC-Pentium microcomputer (Chapleau 1996).

Integrated Systems

Integrating diverse models and methodologies in support of route planning and evaluation of the results can be effectively accomplished in GIS. Choi (1996a) points out that GIS can be used to conflate developed route features into the existing GIS network and how coupling a transportation planning model may generate synergies when conducting economic evaluation. He determined that the integrated system cannot totally replace the route location decision mechanism; rather, it can "shed light on decision making and thereby [generate routes to serve as yardsticks]." He also cautions that "computer systems should not be designed in such a fashion that may allow planners or decision makers to make bad decisions easily." He follows this paper with a report on the use of topological arc-node databases in transit network development (Choi 1997).

Travel Demand Management

Many authorities have chosen to implement travel demand management measures to address some of the negative impacts of transportation in urbanized areas (e.g. air quality and congestion problems). Ridesharing is among the most popular of these measures. While ridesharing can be difficult to implement within a single office or site (unless the site is very large), GIS can be used to manage relevant information from employers throughout an entire metropolitan area. It is more likely that good matches may be found from a larger pool of prospective rideshare commuters. A good match requires proximate locations for home and work, as well as common work schedules. When these attributes are added to other constraints, such as smoking preference or the need to stop by day care, the need for spatial identification of a larger pool of riders is obvious (Zheng 1995). GIS can also facilitate other methods of travel demand management, including:

- parking regulations and reduction of subsidies (GIS to assess parking inventories)
- congestion pricing (GIS to identify travel patterns and socioeconomic characteristics)
- flex time and telecommuting (GIS to assess occupations and industry type)
- on-site child care, assured ride from work to home (GIS to assess socioeconomic and demographic characteristics of commuter sheds)
- provision of secure bicycle storage and shower facilities for bicycle commuters (GIS to identify terrain constraints, bike routes, and existing commuting patterns).

Site Impact Analysis

Traffic impact studies require knowledge of a proposed development's trip generating characteristics, along with spatial information of surrounding streets and land use. Site impact analysis includes assessment of trip generation characteristics, which depend on current and projected land use. It also includes the identification of trip distribution 'splits,' which in turn depend on intervening opportunities and existing travel patterns. GIS is well suited to store the data required by a typical site impact analysis. The analyst can perform innovative studies that more fully consider the relation between land use and transportation. Important features of GIS are its ability to effectively display and report the results of a traffic impact analysis. Examples have included traffic impact studies at airports (Barlett and Bruno 1994) and analyses of hotel-casino trip generation that consider proximity of synergistic pedestrian facilities and land use (Souleyrette et al. 1992).

Network Models

Urban transportation planners employ traditional, four-step models to translate indicators of travel demand to trips assigned to paths on highway and transit networks. These models support medium and long-term decision making, resulting in the development of transportation improvements programs (TIPs) and long-range plan elements. Peterson (1993) suggests that information technology, including GIS, is the key to managing a 30-year transportation plan.

The four-step models are data-intensive, and most commercial packages have typically been difficult to use and have limited graphical interfaces. Further, plotting and presentation tools have been quite limited. Their analytical capabilities, while questioned by some, are powerful and have found widespread use in most urbanized areas of the country. GISs on the other hand, are typically limited in their off-the-shelf transportation analytical ability (Transcad by Caliper Corp. is a notable exception), but have superior data processing and presentation capabilities. As most planning agencies have invested in model packages and many have also invested somewhat in GIS, integration of the two types of systems seems a logical progression. In fact, several efforts have successfully integrated network assignment models with standard GIS (Barrett 1991, Li 1992, Grayson 1993, Kriger et al. 1991, Kriger and Schlosser 1992, TMODEL2 1994, Anderson and Souleyrette 1996). Output from popular transportation planning models can be visualized with GIS.

Still others have used GIS to facilitate various components of the urban planning and network modeling process. For example, Ding (1994) and O'Neill (1991) use GIS to determine optimal design of TAZ. Anjomani et al. (1995) identifies a GIS application for the determination and integration of factors affecting the land use-transportation relationship.

Limitations of conventional sequential, static network modeling are often cited in transportation planning literature. Recently, the Travel Model Improvement

Program (TMIP) has been instituted to address some of the well-known challenges in travel model practice. Among these are developing an understanding of travel as human behavior, developing the ability to accurately model policy decisions, incorporating sensitivity to changes in transit or other modes, and applying technology to transportation. Cervenka (1997) addresses the use of GIS to support person-level travel micro-simulations on a regional scale that may help identify the functional requirements and level of detail for 'a truly effective GIS.'

Freight Transportation Planning

Although not often seen as an urban issue, freight transportation planning also has provided examples of GIS applications in metropolitan areas. For instance, Frazier et al. (1993) described a methodology using GIS to determine truck mileage within a metropolitan planning organization. The technique can be applied to estimate the impact of intermodal transportation on metropolitan truck mileage and hence air quality and congestion. It uses standard GIS software functionality and datasets, including waybill data and U.S. Economic Census information.

In a more recent effort, Souleyrette et al. (1998) use a GIS to develop models in support of a freight planning typology. The typology identifies freight planning issues and advocates analysis of these issues one commodity group at a time. GIS allows summing the results of individual models to provide a picture of aggregate freight demand and flows.

GIS can also be used to inventory and plan safe and efficient truck routes through urban areas. When detailed O-D data are available, commodity type (e.g. hazardous cargo) can be taken into consideration to plan these routes. GIS can be used to store the large, spatial datasets, and its routing tools can be used to determine system-wide least-cost paths. If historical crash information (including rail crossing incident data) is available, it can be tied to network features and their attributes used in conjunction with commodity type to plan 'safest' routes.

TRAFFIC ENGINEERING APPLICATIONS

Application of GIS to traffic engineering is a relatively recent trend. Examples include traffic control, safety analysis, inventory management, impact analysis, emergency response and incident management, and intelligent transportation systems. While limited in its ability to facilitate localized traffic engineering analyses, GIS is well suited to maintain, analyze, and present city-wide spatial data useful in traffic engineering programs. The ability of GIS to integrate advanced technologies such as GPS, surveillance cameras, and vehicle location devices is now being applied to address traffic systems issues in large, congested urbanized areas (Bihn and Gupta 1993). The following sections describe some of the ways GIS is being used in traffic engineering.

Traffic Control and Inventory

Traffic control

Traditional objectives of traffic control are to improve safety and efficiency (reduce delays) in an urban street network. Recently, these objectives have been expanded to include improvements in energy efficiency and air quality. In larger urbanized areas, traffic control systems couple advanced monitoring technology and advanced computer control systems to meet these objectives. GIS provides a method to integrate these technologies by collecting and categorizing spatial information and providing an intuitive user interface for interpretation and analysis of traffic information (Insignares 1991). GIS can also be used to combine data from the system with those from other agencies, "facilitating inter-agency analysis for the management of traffic control problems."

Brooks describes the development of an intersection modeling system (IMS) within GIS that manages data on signalization, geometrics, and traffic volumes for intersections within an urban area (Brooks and Waugh 1994). The system presents data in the form of a digital interactive engineering plan sheet and is able to generate such products as warrant analyses and hourly capacity analyses. The IMS is designed for integration with standard GIS packages.

Sign Management and Signal Inventory

Most urban areas own and maintain thousands of traffic control devices, including signs, signals, and markings. GIS is often cited as the most practical and efficient approach for the collection, inventory, maintenance, and management of these devices. GIS can be used to manage traffic control data and assist engineers in identifying the needs for new signs. It has been used to create lists and maps to support the replacement and maintenance of existing signs (Cipolloni 1992). Such data can be collected in several ways. The use of GPS may be used in some instance to lower total data collection costs (Poling 1993). Some vendors have standards such as the Manual of Uniform Traffic Control Devices (MUTCD) built into their software, which facilitates entry of sign attribute information (CarteGraph 1999). Video may also be captured and integrated in a GIS environment (video log information). With digital and laser technology, high quality images of transportation infrastructure and features can be quickly referenced, via a spatial graphical user interface (GUI), and retrieved with such a system (Bannura 1994).

Safety Analysis

Highway

Several agencies have begun to implement GIS-based crash location and analysis systems. For example, the County of Riverside geographic information system based accident records system (GIS-BARS) utilizes GIS and GPS technology to convert crash data from their native format and develop point topology (Filian and Higelin 1995). Crash data are integrated with centerline roadway attributes and signal information to foster improved safety analyses, using routing, dynamic segmentation, statistics, and buffering functions of GIS. GPS and video-log technologies are used for vehicle location and accident reconstruction applications. Similar systems have been or are in the process of development in North Carolina (Miller 1995) and Iowa (Souleyrette et al. 1998), among others. Choi (1996a) describes an urban application of a crash GIS integrated with a statistical package for analysis of causal factors. Neumann et al. (1996) developed automated geo-coding routines for crash location and analysis using street addresses, milepost information, and offsets from intersections. Recently, developers have also integrated GIS and collision diagram software, creating another powerful analytical tool for engineers and decision makers (e.g. Intersection Magic, PD Programming, Lafayette, Colorado and Accident Information Management System, JNW Engineering, Fairfax, VA).

Rail-Highway

"Every 90 minutes, a vehicle and a train collide at one of the United States' 290,000 highway-rail grade crossings. In 1994, more than 600 people died and more than 1,900 were injured as a result of such crashes" (NCDT 1998). Grade crossing consolidation, especially in urbanized areas, is cited as an effective approach to reducing these injuries and deaths. Agencies such as the North Carolina Department of Transportation have begun to create GIS-based inventory systems to provide field information on each crossing along specified rail routes. GIS can be used to assess the 'cost' of consolidation, including the adverse distance (excess travel time) required after consolidation.

Pedestrian/Bicycle

GIS has also been used to plan safe routes to school. In Clark County (Las Vegas), Nevada, Rasas et al. (1997a) developed maps for each of the county's elementary schools depicting key roadway attributes in areas considered to be within walking distance. Attributes included posted speed limits, traffic signs, signals, crosswalks, and locations of crossing guards. Maps were then provided to schoolchildren, enabling parents to determine safe routes to school. This application utilized GIS thematic mapping and routing functionality.

The effectiveness of child safety seat programs has been analyzed using GIS's and graphically depicted variation in safety seat usage along with zip code socio-economics (Rasas et al. 1997b). This information supports the development of target areas for educational programs or enforcement efforts.

Intelligent Transportation Systems

The relationship between intelligent transportation systems (ITS) and GIS is increasingly being recognized for its importance (APWA 1998, Li 1992). The National ITS Program Plan identifies 29 user services bundled into several primary areas of ITS (ITS 1998a). Some of these areas, with GIS implications, are discussed below, including in-vehicle navigation systems, emergency response and incident management, commercial vehicle operations, and traffic management. It is possible that other ITS applications (e.g. electronic payment and advanced vehicle control and safety systems) could make use of GIS in the future. ITS technology may also be used as an input to static and dynamic GIS databases (e.g. probe cars with GPS and radio links).

In-Vehicle Navigation Systems

In-vehicle navigation systems rely upon maps generated by geographic information systems. Data currency, which is important in all ITS applications, is especially critical in navigation. Since underlying databases are large, static maps are generally provided on CD-ROM or other high density media to drivers, but static maps limit the ability to provide roadway data that may change on a daily basis (e.g. construction zones, new streets, road closures). Furthermore, if real-time traffic conditions are to be provided to road users, there must be a radio link or changeable message sign that provides condition information (Abdel-Aty 1997). Conventional GIS software is incapable of integrating changes and must replace entire maps if any one element is changed. Recent developments, especially in object-oriented GIS, should provide a more efficient way to update data needed for in-vehicle navigation. Improved communications throughput will also facilitate the use of GIS in ITS.

Emergency Response and Incident Management

This application requires the highest level of currency and accuracy in spatial data. Most large urban areas have high quality base maps with comprehensive, unique address ranges and street names. Many rural areas are developing street names and address ranges to support E911 systems. GIS is used in facility planning using location/allocation capabilities. GIS is used in routing vehicles and even in placing vehicles in the most appropriate locations in anticipation of events. Chapter 11 addresses a GIS application in emergency response (disaster management).

Commercial Vehicle Operations

Commercial vehicle operations use GIS maps to support freight mobility in three general areas: fleet management, customer service, and international border crossing passage. Fleet management includes monitoring location, load condition, and vehicle condition, finding stolen vehicles, avoiding traffic congestion, and providing emergency support. Customer service includes providing shippers information on their shipments' locations and estimated arrival times. International border crossing passage includes tracking vehicles to ensure that they have not picked up illegal cargo. GIS can also be used by public agencies to determine tax allocations or user fees, and GIS-based systems are being developed for permitting and scheduling oversize, overweight, and hazardous cargo trucks.

Traffic Management

Some transportation planning and traffic engineering GIS applications may be considered ITS/GIS applications if implemented in real time. An example is the Sacramento Real-Time Ridesharing project, which used a GIS to provide single-trip and multiple-trip real-time ridesharing information. Drivers enter requests into the system and are matched with prospective riders in real time (ITS 1998b).

Implementation Issues

The key to the development of ITS is the implementation of standards to support the national ITS architecture (Okunieff 1995). Regarding spatial information, datum, feature categories, and real-time communications are being considered in the development of spatial data standards. As of this date, the GIS-T community has not developed spatial standards for linear data, although the Federal Geographic Data Committee (FGDC) has developed standards for two-dimensional (2-D) and three-dimensional (3-D) data. Because ITS spatial data are a subset of all GIS-T data, it is important that all-inclusive, easy to utilize standards be developed. The Geographic Data File is the proposed standard for ITS mapping data communications (ITS 1998c).

A cautionary note is provided by Alpert and Haynes (1995), who address the social aspects of the use of GIS in ITS, as well as the use of ITS-generated data for other GIS applications. Privacy implications of each activity were addressed. The authors conclude that GIS applications have a reach that extends well beyond ITS applications (business locations and marketing areas, geodemographics, Census data, financial service data, and health data). Because GIS can be applied in so many different contexts, it "may be difficult to institute meaningful rules or remedies." It is relatively easy to identify individuals with combinations of disaggregate spatial databases and O-D information.

CONCLUSIONS

For some time now, geographic information systems have played an increasingly important role in urban transportation planning and development. In contrast to file-card and pin-map based planning and traffic engineering applications, the modern systems allow for near real-time assessment of conditions along with sharing very large datasets across multiple agencies simultaneously.

Whereas in the past the term GIS has been used to mean "high-tech, expensive mapping and information efforts," many GIS applications no longer bear the stigma of new technology and are accepted and integrated into routine workflow. Where labor efficiency (replacement of labor-intensive methods) was a primary reason for implementation of GIS in natural resources and other fields, reduction of liability and integration of vast, dynamic data streams have been the motivations for use of the same systems in the urban transportation context.

Whether called intelligent vehicle highway systems (IVHS), ITS, or whatever 'fundable' name is chosen in the future, 'smart highways' will rely upon GIS for their underlying spatial base on which to make decisions. And, at the same time, these advanced technologies will produce a great quantity of data, which can best be efficiently stored and retrieved using GIS technology.

REFERENCES

Abdel-Aty, M.A., Abdallah, M.N., and As-Saidi, A.H. (1997). "A methodology for route selection and guidance using GIS and computer network models." *Proceedings, 76th annual meeting of the Transportation Research Board,* Washington, D.C., Paper no. 970016.

Allen, W.G., Mukundan, S., and Faris, J.M. (1993). "Use of GIS in transit alternatives analysis." *4th National Conference on Transportation Planning Methods Applications.* Vol. I and II, A Compendium of Papers, Florida, 3-7.

Alpert, S., and Haynes, K.E. (1995). "Privacy and the intersection of ggeographical information and intelligent transportation systems." *Proceedings of the Conference on Law and Information Policy for Spatial Databases,* Orono, National Center for Geographic Information and Analysis (NCGIA), 198-211.

American Public Works Association. (1998). "Integrating GIS and intelligent transportation systems." Satellite Video Conference, American Public Works Association, Kansas City, MO. http://www.apwa.net/education/dec98.htm

Anderson, M. D. and Souleyrette, R.R. (1996). "A GIS-based transportation forecast model for use in smaller urban and rural areas." *Transportation Research Record 1551,* Transportation Research Board, Washington, D.C., 95-104.

Anjomani, A., Moon, J.H., and Shad, N.A. (1995). "Land use and transportation analysis and integration of determinant factors." *URISA Annual Conference Proceedings,* Washington, D.C., Urban and Regional Information Systems Association, 754-764.

Antonisse, R. (1991). "GIS-T applications in transit: recent experience in Seattle and Boston." *Proceedings of the GIS-T Symposium,* AASHTO, 241-255.

Azar, K.T., and Ferriera, J. (1992). "Visualizing transit demand for current and proposed transit routes." *Proceedings of the GIS-T Symposium*, AASHTO, 133-146.

Bailey, M. and Lewis, S. (1992). "Creating a municipal geographic information system for transportation, case study of Newton Massachusetts," *Transportation Research Record 1364*, Transportation Research Board, Washington, D.C., 113-121.

Bannura, R.K. (1994). 'The role of imaging systems in a GIS-T environment." *Proceedings of the GIS-T Symposium*, AASHTO, 144-150.

Baran, J.F., Lewis, S., and Sutton, J. (1993). "GIS-T developments in an MPO: a case study of SEMCOG." *Proceedings of the Geographic Information Systems for Transportation (GIS-T) Symposium*. American Association of State Highway and Transportation Officials (AASHTO), 314-326.

Barlett, R. E. and Bruno, L.S. (1994). "Use of GIS-T in traffic impact analysis for a new international airport." *Compendium of Technical Papers*, ITE, Dallas, Texas, 394-398.

Barrett, W. (1991). "Building a regional transportation model with GIS software." *GIS/LIS '91 Proceedings*, Atlanta, ACSM-ASPRS-URISA-AM/FM, 80-89.

Bihn, L.W. and Gupta, J.D. (1993). "Snapshot analysis of urban traffic congestion using GIS." *Compendium of Technical Papers*, ITE, 63rd Annual Meeting, The Hague, Netherlands, 357-361.

Brooks, D, and Waugh, M. (1994). "Interactive intersection modeling system within a GIS framework." *Proceedings of the GIS-T Symposium*, AASHTO, 307-327.

CartéGraph Systems, Inc (1999). SIGNview software description, CartéGraph Systems web site, http://www.cartegraph.com/signview.htm.

Cervenka, K. (1997). "Travel sicrosimulation and GIS: an MPO perspective." *Proceedings of the GIS-T Symposium*, AASHTO, 134-140.

Chapleau, R., Allard, B., and Trepanier, M. (1996). "Transit path calculation supported by a special GIS-transit information system." *Transportation Research Record* 1521, Transportation Research Board, Washington, D.C., 98-110.

Choi, K. (1996a). "An expert spatial decision support system for optimum route planning and economic feasibility." *Proceedings of the GIS-T Symposium*, AASHTO, 231-243.

Choi, K. (1996b). "A development of traffic safety investigation tool coupled with GIS and statistical analysis." *Proceedings of the GIS-T Symposium*, AASHTO, 352-362.

Choi, K. (1997). "Transit network development using topological arc-node database." *Proceedings of the GIS-T Symposium*, AASHTO, 357-364.

Cipolloni, M.J. (1992). "Traffic control sign GIS." *Proceedings of the GIS-T Symposium*, AASHTO, 157-172.

Denno, N. and Brail, R.K. (1993). "Improving organizational efficiency through decision support systems for transportation planning." *URISA Annual*

Conference Proceedings, Atlanta, URISA, 2, 94-101.

DeWitt, W. J. and Ralston, B.A. (1996). "GIS: existing and potential applications for logistics and transportation." *Proceedings of the Business Geographics for Educators and Researchers Conference,* Association of American Geographers and GIS World, Inc., Chicago, Ill.

Ding, C. (1994). "Impact analysis of spatial data aggregation on transportation forecasted demand: a GIS approach." *URISA Annual Conference Proceedings,* Washington, D.C, URISA, 362-375.

ESRI (1999). Environmental Systems Research Institute website. http://www.esri.com.

Filian, R. and Higelin, J. (1995). "Traffic engineering in a GIS environment: highlighting progress of the County of Riverside geographic information system based accident records system (GIS-BARS)." *Proceedings of the 15th Annual ESRI User Conference,* URISA.

Frazier, C., Little, P., and Aeppli, A. (1993). "Analysis of truck mileage within an MPO: does intermodal goods movement make a difference?" *Proceedings of the GIS-T Symposium,* AASHTO, 221-232.

Grayson, T.H. (1993). "Transportation modeling using a linked geographic information system and relational database management system." *URISA Annual Conference Proceedings,* Atlanta, URISA, 147-158.

Insignares, M. (1991). "Geographic information systems in traffic control." *Proceedings of the GIS-T Symposium,* AASHTO, 205-214.

ITS (1998a). ITS America Home Page, Resources - Program Plan. Internet: http://www.itsa.org/

ITS (1998b). ITS Project Book January 1997. Internet: http://www.itsa.org/

ITS (1998c). ITS America Home Page, Standards Catalog. Internet: http://www.itsa.org/

Kriger, D., Hossack, M., and Schlosser, M. (1991). "Integration of GIS with the transportation demand model TModel 2 at Saskatchewan Department of Highways and Transportation." *Proceedings of the GIS-T Symposium,* AASHTO, 127-138.

Kriger, D. and Schlosser, M. (1992). "Integration of GIS with a travel demand forecasting model for transportation planning." *Transportation Forum,* 4, 106-115.

Li, Y. (1992). "Using Transcad data to establish an ARC/INFO transportation geographic information system for the Charlotte metro region." *GIS/LIS Proceedings,* San Jose, ACSM-ASPRS-URISA-AM/FM, 2, 459-462.

Miller, S., Johnson, T., Smith, S., and Raymond, L. (1995). "Design and development of a crash referencing and analysis system." *Proceedings of the GIS-T Symposium,* AASHTO, 451-469.

Murakami, E. (1997). "Using GPS for measuring household travel in private vehicles." *Proceedings of the GIS-T Symposium,* AASHTO, 38-48.

Neumann, E.S., Rasas, J., and Subramani, S. (1996). *"Crash statistics for Clark County 1991-1995."* UNLV/TRC/RR-96/05, Transportation Research Center, UNLV, Las Vegas, NV.

North Carolina Department of Transportation. (1998). Rail Division. Internet: http://www.bytrain.org/

Okunieff, P. (1995). "Spatial data standards for advanced public transportation systems - the APTS map and spatial database user requirements document." *URISA Annual Conference Proceedings*, URISA, Washington, D.C., 558-570.

O'Neill, W.A. (1991). "Developing optimal transportation analysis zones using GIS." *Proceedings of the GIS-T Symposium*, AASHTO, 107-115.

Peterson, N. (1993). "Information technology: the key to managing a 30-year transportation plan." *Proceedings of the 21st Annual AURISA Conference*, Adelaide, AURISA, 1-7.

Poling, A., Lee, J., Gregerson, P., and Handly, P. (1993). "Comparison of two sign inventory data collection techniques for GIS." *Proceedings of the GIS-T Symposium*. AASHTO, 179-190.

Rasas, J.P, Fulmer, R.L., and Sathisan, S.K. (1997a). *"Safe route to school - 1997 update."* UNLV/TRC/RR-97/11, Transportation Research Center, UNLV, Las Vegas, NV.

Rasas, J.P, Neumann, E., and Breen, E. (1997b). *"Child safety seat survey."* UNLV/TRC/RR-97/13, Transportation Research Center, UNLV, Las Vegas, NV.

Sirota, S. and Henry, V. (1995). "Using GIS to identify locations with the greatest potential increased light rail ridership.*" Proceedings of the GIS-T Symposium*, AASHTO, 371-382.

Souleyrette, R.R., Maze, T.H., Strauss, T., Preissig, D., and Smadi, A.G. (1998). "A freight planning typology." *Proceedings of the Annual Meeting of the Transportation Research Board*, Washington D.C., paper number 981508.

Souleyrette, R.R., Sathisan, S.K., and Parentela, E.M. (1992). "Hotel-casino trip generation analysis using GIS." *Proceedings, Conference on Site Impact Traffic Assessment: Problems and Solutions*, Robert Paaswell, Editor, American Society of Civil Engineers, Chicago, IL.

Souleyrette, R., Strauss, T., Pawlovich, M., and Estochen, B. (1998). "GIS ALAS: the integration and analysis of highway crash data in a GIS environment." Geographic Information Systems for Transportation Symposium, American Association of State Highway and Transportation Officials, Salt Lake City, Utah, April 20-22, 411-428.

TDMODEL2 (1994). "TMODEL2 linkages to GIS," TMODEL Reports, 4(1). Internet: http://www.tmodel.com/linkag.html.

Zheng, J., Greenfeld, J., and Mouskos, K.C. (1995). "A GIS based ride-sharing advanced traveler information system (RSATIS*)." Proceedings of the GIS-T Symposium*, AASHTO, 383-400.

CHAPTER 8

Public Utilities

Songnian Li, David J. Coleman, and Said Easa

Traditionally, there are three major challenges facing the utility industry: large service areas, many distributed customers, and remotely distributed aging facilities. Recent deregulation and increasing market competition also have imposed more sophisticated difficulties. Utilities are seeking new technologies to tackle these challenges, and automated mapping/facilities management (AM/FM) integrated with emerging technologies, provides vital solutions. In this chapter, current AM/FM applications in public utilities such as water and wastewater, electricity, cable television, telephone, and telecommunication are examined. The relationship between AM/FM systems and some emerging technologies is explored. Emerging developments of AM/FM that help promote more efficient management, customer services, and technical support are described.

INTRODUCTION

Utilities have used 'intelligent computerized mapping' systems, also known as automated mapping/facilities management (AM/FM) or geographic information systems (GIS), since the early 1970's. In simple terms, Antenucci (1991) defines an AM/FM system as "a geographic information system that integrates non-graphic facilities management information into a database that is tied to facilities maps." In most cases, the topological data structures employed by modern AM/FM software also include sufficient information to store and update connectivity relationships between facilities, thus simplifying utility network analysis applications.

The need to attract and keep customers in a changing world is increasingly becoming a more significant reason for using AM/FM in utility applications (Meyers 1999, Fry 1999). Public utilities have many uses of AM/FM, most of which focus on the need to provide cost-effective services to consumers and better management of resources. Companies across utility industries have demonstrably accepted GIS as a necessary tool for their business. When one takes a look at the nature of the information with which utilities deal, it becomes clear that AM/FM can play a critical role for them.

The emergence of new technologies and the increased competitive markets have imposed more requirements on public utility companies. It has become critical for these companies to find a way to improve their competitive positions in terms of management, service, and marketing strategies. AM/FM technologies have proven to help solve these problems for over twenty years.

This chapter first presents an overview of the development of AM/FM systems in public utilities. The discussion will focus on pipeline-based and cable-based systems. Current AM/FM applications in public utilities are then described and the relationship between these applications and newly emerging technologies is explored. Finally, the impacts of emerging developments, such as Internet-based technologies, GIS Toolkits, object-oriented technology, and supervisory control and data acquisition (SCADA) systems are discussed.

DEVELOPMENT OF AM/FM SYSTEMS

Background

Public utility networks comprise the most important and extensive infrastructure in any city, state, or country. They provide essential support for running a society on uninterrupted daily basis, in most countries. The service area of a single utility company could vary from several hundred to many thousands square miles, and may serve anywhere from a few thousand to several million customers. With many distributed facilities remotely located, utility systems become more sophisticated to manage and maintain.

Compared with other businesses, utilities are unique in that: (a) they have distribution networks that must be maintained and (b) the location and selected attribute information on such networks must be shared with other utility providers working in the same areas. Mahoney (1991) characterized the 'utilities industry' as consisting of gas, water, wastewater, and electricity providers. The form and category of public utilities vary among countries due to different historical developments and socioeconomic characteristics. However, for purposes of this review, public utilities are grouped into two categories:

- *pipeline-based systems:* gas, water, and wastewater
- *cable-based systems:* electricity, telecommunications, cable TV, and the Internet.

These two categories have historically been the most powerful users of AM/FM technology. Their applications are diverse and mature, but continually impose new challenges. Although the components of a public utility vary from one utility to another, with respect to facilities and services, utilities do have the following common characteristics:

- All utilities possess a physical network infrastructure with facilities/plants locally or remotely attached. This network must be maintained and its information (especially location data) shared with other utilities that share the same land use.
- They usually provide services in a similar way within a regulatory framework.
- They require similar geographic information to support their operations, including: (1) property/land use maps, (2) locations of pipelines and cable systems, (3) street/road networks, and (4) locations of other municipal facilities.
- They use similar spatial datasets and have similar work order management requirements with three basic processes: production, transmission, and service.
- Their routine operations are supported by the same types of spatial data management, analysis, and output functions, including: (1) load and network

analysis, (2) records keeping and reporting, (3) facility mapping, (4) outage analysis, (5) maintenance and inventory, and (6) market analysis and customer service.

Historical Perspective

Since their inception in the late 19th and early 20th centuries, operating and maintaining distribution services of major utilities have required some degree of records management. Whether these records represented survey plans (covering existing and proposed rights of way), construction plans, facilities layouts, work orders, or customer service records all contained information that could be geographically referenced. In addition, many records contained a combination of graphics and textual information. The requirements for timely and up-to-date records management grew as government regulatory bodies stepped up their scrutiny of the utilities' operations, especially application for rate increases. McDaniel et al. (1998) offers an excellent treatment of the history of AM/FM developments in North America.

Virtually all engineering-related activities and many customer service activities required at least rudimentary basic mapping covering a utility's entire service area. In the U.S., responsibilities for property mapping and basic topographic mapping (at scales larger than 1:24,000) vary widely from state to state. As a result, it was not unusual for utility construction staff, distribution planning engineers, and facility managers to be faced with the need to represent existing or proposed services on portions of source maps from several different sources. This typically involved rationalizing different map scales, coordinate systems, coverage areas, accuracy, and cartographic symbolization systems. In many cases, representing all facilities on a single, consistent base map resulted in considerable cost savings to engineering and maintenance operations. It was the influence of these cost savings on preliminary benefit-cost (B-C) analysis that helped provide the economic justification to proceed with early AM/FM implementations.

By completing and rationalizing this mapping over their respective service areas, the utilities themselves became important sources of base mapping to other users. Further, by relating the mapping to a coordinate system, based on either a local grid or a projection-based state plane coordinate system, utilities were able to relate their facilities to those of other utilities. Early pioneers in this area included Bell Telephone of Pennsylvania, Pennsylvania Power, and Light and Pacific Bell. The effort of the Public Service Company of Colorado from 1968-1972 represents one of the best known early initiatives. Other pioneering implementations included Northern States Power, Texas Power and Light, and San Diego Gas and Electric (McDaniel et al. 1998).

AM/FM technology was adopted by utilities in Canada, the United Kingdom, and much of Western Europe somewhat later than in the U.S. The favorable B-C ratios that justified many American AM/FM initiatives in the 1970's and 1980's, predicated on significant incremental increases in efficiency as a result of creating a common base map for operating areas, could not be matched by utilities in other countries. Generally speaking, due to the existence of different institutional arrangements, more comprehensive medium and large-scale base maps (municipal, state/provincial, or even national) were already available from governments and in use by utilities in each country. The incremental benefits accrued through simple use of a digital version of these base maps were not enough to justify the high cost of hardware, software, data conversion, training, and on-going database maintenance associated with AM/FM implementation.

Sufficient justification was finally made as a result of the tremendous pressures of deregulation and privatization faced by utilities from the mid-1980's onwards. To successfully compete for customers, the utility industry had to develop new strategies and management tools to improve customer service and maintain their asset bases. Many emerging technologies have begun to play important roles for utilities, such as Internet technology, component software technology, object-oriented technology, and SCADA. The actual effects of these emerging technologies on public utilities will be discussed later.

Today's utilities must be concerned with knowing who their customers are, where they are located, what their usage patterns are, and what types of services they prefer. Furthermore, they need to analyze information, model and simulate realty, and make better decisions. AM/FM technology supports this need by means of powerful capabilities for mapping process automation, record keeping, spatial analysis, and data/application integration. In this regard, AM/FM systems have been compelled to move beyond their early roots in 'end-user engineering computing.' As with their GIS counterparts, increasingly sophisticated customer demands and operational requirements have forced AM/FM systems to move from project-based implementations to enterprise-wide deployment (ESRI 1997). This topic will be discussed further in the next section.

APPLICATIONS

AM/FM technology has been applied to many areas in public utilities. These include automated mapping and map maintenance, infrastructure siting, records keeping and management, planning/routing, customer information service, network analysis (free duct, marketing, and operation), work order management, and decision-making. Routinely employed generic functions focus on facilities database inquiry (e.g. engineering and service), utilities maintenance (e.g. plant replacement and records maintenance), analysis (e.g. outage analysis and service area analysis), planning (e.g. network planning and facilities planning), and decision-support optimization (Easa et. al. 1997).

Current AM/FM applications, benefits, costs, limitations, and new developments have been extensively studied and documented in Antenucci et al. (1991), Mahoney (1991), Ray (1996), Sarinas (1998), Toffer (1998), Meyers (1999), and Fry (1999). The same information can also be found in the proceedings of both AM/FM International Conferences (now GITA: Geographic Information Technology Association) and the Urban and Regional Information Systems Association (URISA).

As integrated systems become more powerful and alternative spatial datasets become more widely available, more important value-added applications have been realized, including market analysis, decision support, and work order management.

Now, most utilities are focusing on reducing cost, attracting and keeping customers, and remaining competitive in the market. Under the continuing pressure of deregulation and distribution, utilities are even facing more challenges in using AM/FM to support new applications such as collaborative decision-making and distributed map services.

The following sections discuss some AM/FM applications, grouped into three categories: data management and facilities mapping, operation support using spatial analysis, and enterprise integration.

Data Management and Facilities Mapping

The geography of utility records is defined by: (1) the topographic base map to which facilities records are spatially referenced and (2) the locations of facilities such as pipelines and cables in relation to this base map (Mahoney 1991). Conventionally, infrastructure facilities of public utilities were managed using paper-based maps and records and a variety of text-based computer systems. AM/FM systems have been used to convert these conventional systems into a digital form, integrate other data, maintain the records on-line (in a real time mode) or off-line, and automate mapping processes for engineering design/planning, infrastructure maintaining/updating, and customer service.

Spatial data capture and database creation/maintenance of any existing AM/FM system remain a costly, time-consuming task. New technologies have emerged to facilitate data capture and acquisition both locally and by remote field crew. Acquisition of data through mobile terminals, laser technology, GPS, and SCADA, and incorporation of data into geo-referenced mobile solutions and pen-based computing systems have been reported (Bernard 1994, Daly et. al. 1994, Gay 1994, Kendall 1998, Coleman 1999, Meyers 1999). In addition, the users now have a much richer array of choices regarding base maps and value-added thematic data than in the past (SAIF 1999). This is due to the growing number of available datasets, advances in metadata documentation, and tools for translating digital map files between different systems, like feature manipulation engine (FME).

Typically, GPS has been used in AM/FM applications for establishing basic control points. Recently, however, it has gained wider use in field data collection, mobile positioning, and data maintenance. By integrating GPS with other handheld measurement tools and (in some cases) pen-based computing systems, the acquisition of both spatial and attribute data at the same time has now become feasible (Bernard 1994). A wide range of such systems is available from mainstream and niche-market vendors alike.

Operation Support Using Spatial Analysis

Spatial analysis is the most outstanding capability in AM/FM systems which support facilities design and planning, operations management, and customer service applications. Because the topological nature of utility networks (connectivity relationships in particular) may be modeled using fairly generic and well-developed tools and data structures, network analysis has become one of the 'workhorse applications' underlying AM/FM usage in most utilities. The applications of AM/FM technology in the utility workplace are numerous. Table 1 lists specific applications of spatial analysis capabilities in selected utility sectors. The spatial analysis application areas common to all utilities are:

- engineering planning and design
- network planning, modeling, and simulation
- trouble spots/plants identification
- emergency/service crew dispatching
- market and needs pattern analysis.

Ray (1996) points out that providing high levels of service at minimum cost (that customers demand) can only be achieved by managing assets efficiently and understanding their effects on service quality. In this regard, AM/FM systems have proven their principal worth with regard to basic mapping and inventory functions. However, information integration and modeling capabilities of AM/FM systems make them even useful for strategic planning and support of day-to-day routine

Table 1   Specific spatial analysis applications for various utilities

Utility	Spatial analysis applications
Electricity	Outage analysis (dealing with trouble calls) Transmission line siting Load pattern and growth analysis Impact analysis in facilities siting
Water/wastewater	Breakage and leakage diagnosing Water network flow analysis Modeling damage to water distribution systems Water resources planning and management Simulation of ground water mass destruction Prediction of runoff rates Determination of pressure zone when planning new water distribution facilities
Telephone/cable TV	Network/cable routing Facilities siting and location optimization Outage and performance problem analysis Black spot/zone analysis in cable television
Telecommunications	Radio propagation and area coverage analysis Optimum antenna heights and locations using DTM information Optimal design of a broadband network layout Analyzing tower coverage areas and service accessibility of a mobile communications network Network traffic analysis by combing demography

activities. Strategic planning may include expansion of existing facilities, planning of new services, and siting of new transmission lines.

The AM/FM technology permits the integration of base mapping, ownership information, property and political boundaries, and existing/proposed land use information to identify potential opportunities and constraints to development (Sarinas 1996). Another important application area is work order management that includes issue work orders, dispatch service crews, schedule equipment (e.g. transmission and distribution equipment), inspection programs, and control workflow.

Scattered geographic service areas require work order management systems that identify costs related to specific activities by location. AM/FM helps provide the data necessary to evaluate efficient operations and integrate data into work orders and tasks. In addition, many tools within AM/FM systems provide routine operational support, to other real-time data acquisition applications, for trouble call/outage analysis, distribution automation, customer service, leak detection, maintenance, automated meter reading, and SCADA. For example, Davis (1994) noted three benefits that could be gained by using AM/FM to perform an outage analysis (Figure 1):

- helping operators of a trouble-call system to quickly pinpoint the problems
- providing valuable information online for dispatching trouble crews, such as site maps, routing, and maintenance records of the faulty facilities
- providing vital data for the infrastructure replacements through cumulative analysis of outage records.

Enterprise Integration

As discussed in Coleman (1999) and Popko (1988), large utilities and municipalities were among the first to identify the operational requirements to integrate smaller GIS and FM-related databases on stand-alone systems with large corporate databases residing on mainframes. As local area networks became more widespread, enterprise computer systems evolved from single-tier (host-based) to two-tier environments with networked PC's and workstations in a client/server architecture replacing connected terminals. This modular approach is attractive to many or ganizations because it allows them to take quicker advantage of

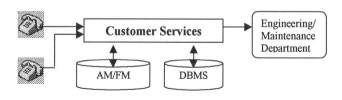

Figure 1 Trouble-call system for outage management

price/performance improvements on modular hardware components as well as the other advantages of network computing described earlier (Mimno 1996).

While advantageous in many respects, these two-tier environments do not necessarily have the advantages of centralized host-based implementations, namely availability, expandability, and reliability of service (Strand 1995). As a result, newer three-tier environments have been developed to aid the placement of applications and data at locations that optimize these three factors (Figure 2).

Existing mainframes at Tier 3, possibly connected over long distances via wide area networks (WAN) or even Internet Service Providers, are used to support legacy applications and provide access to large databases. To implement a client/server computing strategy, additional layers of hardware and software are added as a front end to the host computer. These layers consist of shared servers at Tier 2 (interconnected by high-speed LANs), and LAN-based PCs and laptop computers at Tier 1. GIS and desktop mapping applications are included at Tier 1, although the data may reside on either Tier 2 servers or even Tier 3 hosts.

These 'enterprise information architectures' have proven to be highly flexible to be easily modified to accommodate changes in business requirements. As more customers adopt this approach, commercial GIS software firms and mainstream DBMS vendors alike are modifying their offerings in response to customer demands. In particular, spatial data management models and processes are becoming more intrinsic components of an organization's larger information management architecture. ESRI's *Spatial Database Engine*™ is one commercial example of how GIS vendors are providing even tighter high-performance links into mainstream relational DBMS packages, like Oracle8i, while based on de facto industry standards for object linking and embedding (OLE), SQL, 3rd and 4th generation languages, and ODBC.

Today, it is commonly accepted that AM/FM should become an important part of the larger information technology (IT) solution of any utility. As Levinsohn (1997) argued, "Greater strategic benefits and returns on investment are possible when AM/FM is developed as an enterprise-wide initiative that supports several business functions." However, there is a variety of ways in which this integration may take place (Juhl 1998). Current integration implementations may be classified into three categories:

- integrate other IT systems based on AM/FM
- integrate AM/FM functional modules into other application systems
- embed AM/FM components into the core of the entire enterprise IT system

Regardless of the approach taken, the ultimate success of such integration relies heavily on three issues: the quality and effectiveness of software interfaces among the integrated systems, communication between systems, and the degree to which spatial datasets *in their basic form* can be meaningfully shared among the applications. While mileage may clearly vary, significant benefits may be realized through closer integration of AM/FM applications into the overall enterprise information architecture of a utility. Recent studies have described the benefits and caveats of tighter integration with existing legacy databases, accounting/financing, CAD, work order/asset/document management, support applications, multimedia, and access to other utilities' systems (Ginther and Lopez 1994, Daly and Lebakken 1994, Ayers 1996, Nicoloro 1994, Gay 1994).

IMPLEMENTATION

Public utilities have realized that AM/FM can be a valuable tool to improve efficiency and productivity. It is not only just for facility mapping, but also for better data integration, decision-making, infrastructure management, work order management, and market/customer services. With the use of AM/FM, utility companies also have opportunities to change their industry images and strengthen their ability to compete in the market.

The benefits accrued by using AM/FM technologies in public utilities may be either tangible or intangible. While it is more desirable to emphasize tangible benefits, the initially intangible benefits often deliver the higher return over the long term. This is because tangible benefits are easier to evaluate and more accepted by decision-makers. In justifying either the launch of new AM/FM systems or the extension of existing ones, consultants and internal staff look to quantify the benefits accruing from five key areas (Meyers 1999):

- increased productivity in mapping
- increased productivity through information access
- improved engineering and operations efficiency
- decreased use of mainframe computers
- elimination of costs due to map and record replacement/consolidation.

The intangible benefits in the last area may include:
1. Improved service through the ability to integrate internal information with that from other utilities

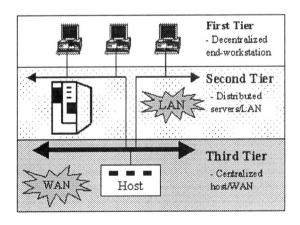

Figure 2   Example of three-tier architecture

2. The ability to develop and offer new value-added products or services through data integration
3. The ability to make better strategic planning decisions due to faster access to a wider variety of integrated datasets
4. Demonstrated growth in customer base through the perception of the utility as a user of leading-edge technology (though this may sound naive).

The AM/FM benefits do not come without a price. Typically, organizations incur high up-front costs associated with initial purchase of hardware and software. Further, any significant financial paybacks through savings and fees typically do not accrue until much of the database has been loaded and initial applications have been developed. Since neither may occur during the first 3-5 years of a program, the discounted B-C ratios for some periods may appear unfavorable to managers unfamiliar with the dynamics of AM/FM investments. As well, even with top-level support for the initiative, building consensus on data sharing at the corporate level can be very time-consuming and may deliver only limited results over the short term.

EMERGING DEVELOPMENTS

Intranet/Extranet/Internet

Since 1993 (in particular), the wired services built through the Internet have defined current paradigms for mass-market network usage. Utilities today are rapidly embracing the World Wide Web (WWW): (1) as a means to provide applications that aid customer service externally and (2) as an alternative to other types of middleware to connect systems and people with their organizations (Black 1997). Intranets/Extranets are those internal, restricted corporate networks that are based on Internet protocols to communicate information and improve relationships with employees, clients, partners, and customers. Public utilities may be web-enabled on the corporate Intranets/Extranets and the Internet. This impressive technology has the potential to produce dramatic operational benefits through improved data communication and collaboration among partners.

Intranet/Extranet-enabled AM/FM applications permit database queries, integrated quality assurance and control, interactive editing and updating of database records, and collaborative workflow for enterprise work management support. Plewe (1997) provides an excellent summary of the capabilities and limitations of Internet-based mapping and GIS software as it existed in mid 1990's. Besides, many web-enabled workflow systems may be integrated with AM/FM to support its routing tasks over Intranet/Extranet models. The city of Kingston, Ontario, Canada, provides a good example of a utility GIS Intranet that uses Autodesk's MapGuide^{TM} technology, including MapGuide^{TM} server on the server side and Netscape Navigator with MapGuide^{TM} plug-ins on the client side (http://www.hunter-gis.com/).

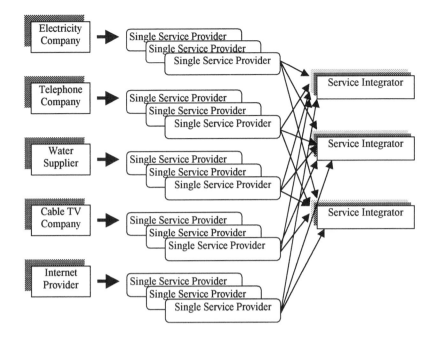

Figure 3   Process of utility deregulation and service integration

AM/FM utility users usually include different groups distributed across different offices at many different locations. For example, some utilities may break down into:

- production groups, each is responsible for generating utility products (e.g. electricity, water, and television programs)
- a wholesale group that is responsible for transmitting the products to the distribution centers, or
- a retailer group that is responsible for door to door sale.

A new scheme has emerged in which a utility service integrator may offer residential consumers electricity, communications, natural gas, water, cable television, newspaper, and Internet service through a single point contact (Juhl 1998). Figure 3 depicts this scheme which clearly adds more complexity to the distributed nature of utilities. As illustrated, a 'single service provider' provides solo utility service, while a 'service integrator' provides versatile utility services.

By using Intranet/Extranet-based AM/FM systems, the information flows and databases within a utility can be updated in real time, and collaborations among different groups can be made more effective. These collaborations allow easy transfers of large-size digital mapping files, enable group editing and decision making based on graphics data, and easily provide different views of distributed (maybe remotely located) map contents. Some useful information can be found in Li et al. (1998).

Internet-based AM/FM approaches are still relatively new and are constrained by the lack of appropriately reorganized databases, current limited capabilities of WWW protocols, and relatively slow performance of such systems over standard communication lines (Coleman and Nebert 1998). Other technologies like mobile computing, networking, wireless communications, and light client must also be properly addressed in terms of implementing Internet-based AM/FM systems. However, current developments with Java-based web applications and new WWW-based products promise to provide users with more transparent access to distributed databases and datasets created on a variety of software platforms.

Component Software Technology (GIS Toolkits)

AM/FM software packages were originally developed with a strong emphasis on providing end-users with a full range of functions and embedded procedures. Such tools are necessary to collect, edit, manage, analyze, and display spatial data and its attributes in a wide range of applications. Some systems allowed customization to permit individual user groups to use selected capabilities for specific purposes. However, given the relatively narrow niche markets involved, software vendors were under increasing pressure to provide versatile products to a wider market. The result was that most popular and best-selling packages became very large and complex. Learning curves steeped, documentation became intimidating, and even resident software experts within a customer organization began focussing on only certain software aspects that are required for a particular set of applications.

At the same time, a number of commercial database management system (DBMS) software products (e.g. Oracle and Informix) began to incorporate more complex spatial data management functions and data structures within their mainstream product lines. These new software releases did not posses all the specialized spatial data handling and custom application functions inherent in AM/FM packages. However, the mainstream DBMS firms did possess the advantage of a long tradition of managing huge attribute databases maintained by utilities, municipalities, and other large organizations.

During the past five years, vendors have launched new products that respond to the problem of steeping learning curves of mainstream AM/FM software and the emerging 'spatially aware' DBMS. New object-based GIS-toolkits like MapObjects^{TM} from ESRI, MapX^{TM} from MapInfo Corporation, and CARIS++ Framework^{TM} from Universal Systems are designed to provide users with a collection of very focused spatial functions/components that could individually or collectively interact with the data stored in these larger commercial databases. These spatial functions/components can be incorporated to aid existing applications, build lightweight data viewing applications, and create new AM/FM programs to solve specific tasks.

Using languages like Visual Basic, Delphi, C++, or Java, developers can now create custom spatial applications that meet the immediate needs of the organization without possessing the complexity of the larger mainstream packages. Some examples are:

1. US-based Air Touch Cellular Inc. will use Universal's CARIS++ Framework to create a software package for planning and analyzing transmission towers for expanding U.S. cellular phone market.
2. Ericsson-Hewlett Packard Telecommunications AB will use CARIS++ Framework to create a tool for modeling and mapping in support of service planning and radio tower location.
3. Microsoft Corporation has employed MapX in its integration of limited desktop mapping capabilities into the MS Excel software.
4. Elsinore Valley Municipal Water District used MapObjects to create a District-wide 'front end' to the EVMWD GIS database. This interface allows doing spatial query quickly and easily. It also provides access to an interactive, dynamic map of the entire District and infrastructure (Ollerton 1998).
5. Jefferson County GIS Department used MapObjects together with other technologies like Sylvan OCX tools, Visual Basic, and AML to develop an enterprise front-end interface known as the data extraction interface (DEI). The interface is used to serve 2500 employees, among them 1600 have PCs. The benefits of using MapObjects include faster performance, streamlined and standardized development, and multiple architectures, while licensing cost and development time are being identified as disadvantages (Gallaher 1998).

These toolkits promise to offer users the sophisticated functionality they require with a smaller investment in training and long-term software support. Although these toolkits will not replace the overall need for mainstream GIS products, there will be an increasing proliferation of such packages in the marketplace as the demand for 'spatially aware' applications increases. Since GIS developers are already using their own toolkits to build the next generation of their products as well, the line between shrink-wrapped products from these vendors and the toolkits themselves will become blurred.

SCADA (Real-Time System)

Supervisory control and data acquisition (SCADA) is the process by which real-time information is gathered from remote locations (for processing and analysis) and equipment is controlled. SCADA is fundamental to any utility operation that involves observation and manipulation of remotely located processes. It is used in the electric, telecommunications, pipeline, water and wastewater, oil and gas, and infrastructure and transportation fields, where its primary function is to continuously monitor and control critical parameters throughout the network from a central location. For example, power transmission and distribution systems use SCADA to monitor and control the delivery of power from the generation sites to customers (Daly and Lebakken 1994). Water systems may use SCADA technology to implement a real smart metering system to record the amount of water consumed. The parameters controlled include flow data and performance data such as pressure and temperature.

A SCADA system typically consists of a master terminal unit (MTU) and one or more remote terminal units (RTU) plus supporting software (Rhodes 1994, Nicoloro 1994). A typical master unit consists of a base processing unit, operator interface, power supply, modem, input/output cards, relays, enclosure, battery backup, wiring terminals, lightning protection, programs, and start-up. A typical remote unit consists of a base recessing (instead of processing) unit and all the preceding components, except operator interface and programs. Other accessories can include radios, alarm dialers, field instruments, computers, and printers.

SCADA systems data have been routed on from traditional leased lines to not only over dedicated fiber, microwave, or 900 MHz multiple-address systems, but also over high-bandwidth corporate communications networks (Rhodes 1994). Whilst almost all SCADA systems are centered on a computer system, the architecture can be either centralized or distributed (Gay 1994).

There are two major uses of SCADA systems in public utilities. One is providing data from key points in a network to help anticipate problems in gas, electric, and water delivery systems and identifying the nature and location of a problems in all networks (Meyers 1999). Another is to meter customer usage of service and control the service status. By integrating SCADA and AM/FM, this information can be transmitted into the AM/FM system and associated with spatial data to facilitate further processing. 'Live' maps and real-time databases that are used to manage large utility systems become feasible. Through integration, SCADA becomes spatially related and AM/FM systems become real-time. This means that real-time data from multiple distribution automation and SCADA systems can be imported and graphically displayed over the geographical background provided by AM/FM systems (Daly and Lebakken 1994). For example, SCADA used in gas pipeline companies is represented as a large wall model of the regional pipeline network. The model contains real-time information on gas flow, pipeline pressure, sections under repair, alternative pipeline routings, location and dispatch of service crews, and so on.

Some limitations exist in integrating AM/FM and SCADA, relating to system architecture, information retrieval, and data access. The real-time link between AM/FM and SCADA systems is limited and depends on the architecture of the integrated system (Daly and Lebakken 1994). Ayers (1996) stated several options for the linkage (Figure 4). However, none of these options is perfect because in handling temporal data generated from time-based systems, we still need to

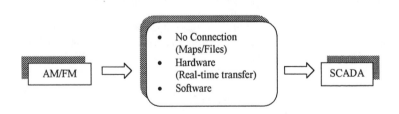

Figure 4   Approaches to linking AM/FM and SCADA

balance data integrity and speed, storage and time, and so on. Also, the problems of accessing the data generated from time-based systems are not well solved. No good solution has been found for users to directly access the data without knowing technical details. Ayers also addressed other issues related to using SCADA such as writing data to RDBMS, routing data directly through WAN, and creating specific data handling applications.

CONCLUSIONS

This chapter has described the application areas and implementation issues of AM/FM in public utilities and discussed emerging developments. Rather than covering the wide array of current and emerging technologies and applications, the chapter has focused instead on several key issues and applications and referred the reader to relevant sources for more details.

As mentioned earlier, utilities in North America and Western Europe were leaders in the development, implementation, and refinement of AM/FM systems and concepts. As with their counterparts in natural resources, these systems grew in environments characterized (and usually led) by strong-willed end-users of the technology. Management information system (MIS) groups within the utility industry are subsuming many of these systems. Especially with the release of spatial enhancements to commercial systems (e.g. Oracle8i's Spatial Data Cartridge), the AM/FM software's spatial functionality is becoming submerged within larger mainstream database management systems. There is an increasing belief that the concepts, traditions, practices, and even lessons learned in AM/FM over the past twenty years will be lost.

It is true that much of the mechanics of the processes must be surrendered to application programmers and system analysts as the technology becomes more complex. However, it is the engineers and end-users that still drive the requirement definition and acceptance testing aspects of the technology. These aspects must not be relegated to MIS staff, even if the staff is competent. To be truly responsive to the needs of operations managers and customers alike, *teams* of end-users, senior engineers, and developers must ultimately take responsibility for the definition of further developments and the extensions to existing packages.

REFERENCES

Antenucci, J.K., Brown, K. Croswell, P.L., and Kevany, M.J. (1991). *Geographic information systems: a guide to the technology*, Van Nostrand Reinhold, New York, N.Y.

Ayers, L.M. (1996). "AM/FM/SCADA integration issues: a technology primer." White Paper, PlanGraphics Inc., Redmond, WA.

Batty, P.M. and Bundock, M. (1996). "The impact of new technologies on AM/FM." *Proceedings of AURISA '96 Conference*, Hobart, Australia.

Bernard, J.A. (1994). "Attribute collection and GPS." *Proceedings of AM/FM '94 Conference XVII*, 169-175.

Black, J.D. (1997). "Utilities move towards transparent GIS." *GIS World*, 10(2), 46-48.

Coleman, D.J. (1999). "Geographic information systems in networked environments." In *Geographical Information Systems*, 2nd Edition, D. J. Maguire, M. Goodchild, D. Rhind and P. Longley (eds.), John Wiley & Sons, London, United Kingdom.

Coleman, D.J. and Nebert, D.D. (1998). "Building a North American spatial data infrastructure" *Cartography and Geographic Information Systems*, 25(3), 151-160.

Daly, P.A. and Lebakken, T.M. (1994). "Above and beyond the barriers to AM/FM/SCADA integration: lessons learned from the EPRI-NRECA Pierce-Pepin project." *Proceedings of AM/FM '94 Conference XVII*, 303-316.

Davis, G.L. (1994). "Mapping in support of outage management: Southern California Edison's approach." *Proceedings of AM/FM '94 Conference XVII*, 757-767.

Easa, S.M., Li, S., and Hamed, M.M. (1997) "GIS applications in infrastructure management: state of the art." *Proceedings of the International Conference on Rehabilitation and Development of Civil Engineering Infrastructure Systems*, June 9-11, Beirut, Lebanon.

ESRI (1995). "ARC/INFO prototyping AM/FM applications: quality/schedule tradeoffs." *White Paper*, ESRI, Redlands, CA.

ESRI (1997). "GIS data storage trends: implementations for utilities." *White Paper*, ESRI, Redlands, CA.

Fernandez, R.B. (1994). "SCADA data integration with facilities management in Miami-Dade." *Proceedings of AM/FM '94 Conference XVII*, 149-159.

Fry, C. (1999). "GIS in telecommunications." *Geographical Information Systems: Management Issues and Applications*, John Wiley & Sons, New York, N.Y.

Gallaher, D.W., Massaro, N., and Mitchell, S. (1998). "Enterprise GIS: a cost-effective approach." *ESRI User Conference Proceedings*, San Diego, CA.

Gay, R. (1994). "Integrating GIS and SCADA systems." *Proceedings of AM/FM '94 Conference XVII*, 581-587.

Ginther, P. and Lopez, T. (1994). "Integrating pipeline applications through AM/FM." *Proceedings of AM/FM '94 Conference XVII*, 461-472.

Juhl, G. (1998). "Utilities race for a competitive edge." *GIS World*, 11(5), 43-46.

Kendall, T.G. (1998). "Integration ignites state-of-the-art AM/FM." *GIS World*, 11(5), 48-50.

Levinsohn, A. (1997). "AM/FM vital to evolving utility industry." *GIS World*, 10(12), 38.

Li, S., Boettcher, R., and Coleman, D.J. (1998). "Geomatics workflow management requirements in a collaborative production environment." *Proceedings of the American Congress on Surveying and Mapping Conference*, Baltimore, 269.

Mahoney, R.P. (1991). "GIS and utilities." *Geographic Information Systems: Principles and Applications*, Longman Group UK Limited, London.

McDaniel, K., Howard, C., and Emery, H. (1998). "AM/FM entry: 25-year history." *History of Geographic Information Systems*, Prentice Hall, Inc., New York, N.Y., 119-146.

McLeroy, R.J. (1994). "Linking AM/FM with work order management and hydraulic analysis models." *Proceedings of AM/FM'94 Conference XVII*, Denver, USA, 373-383.

Merritt, S.D. (1997). "AM/FM makes business sense." *GIS World*, 10(1), 66.

Meyers, J. R. (1999). "GIS in the utilities." *Geographical Information Systems: Management Issues and Applications*, John Wiley & Sons, New York, N.Y., 801-818.

Mimno, P. (1996) "Building enterprise-class client/server applications". White Paper. Dynasty Technologies, Inc., Lisle, IL, USA. Internet: *http://www. dynasty.com/product/mimno_wp.htm*.

Nicoloro, M.A. (1994). "Commonwealth gas company SCADA/Gass project implementation and future linkage to AM/FM." *Proceedings of AM/FM'94 Conference XVII*, Denver, CO, 59-68.

Ollerton, J.A. (1998). "Making GIS available using ESRI MapObjects technology: A study of Elsinore Valley Municipal Water District's MapObjects implementation." *ESRI User Conference Proceedings*, San Diego, CA.

Owen, P.K. (1998). "Teleco GIS proves its worth." *GIS World*, 11(7), 44-46.

Plewe, B. (1997). *GIS ONLINE: information retrieval, mapping, and the Internet*, ONWORD Press, Santa Fe, NM87505-4835, USA.

Popko, E. (1988). "An enterprise implementation of AM/FM." Internal Paper, Geo-Facilities Information Systems Application Center, IBM Corporation, Houston, Texas.

Ray, C.F. (1996). "The use of GIS in a major water utility company geographic information systems." *Proceedings of the Institution of Civil Engineers*, 114(2), 23-29.

Rhodes, R.A. (1994). "Making SCADA a sustainable competitive advantage." *Proceedings of AM/FM'94 Conference XVII*, Denver, CO, 225-230.

SAIF (1999). "The spatial archive and interchange format." *FMEBC home page*, http://www.env.gov.bc.ca/gdbc/fmebc/SAIF_FMEBC.htm.

Sarinas, E. (1996). "Implementing a GIS for Puget Sound power and light." http://weber.u.washington.edu/~immature/207/casest.html.

Strand, T.M. (1996). "GIS thrives in three-tier enterprise environments." *GIS World*, 9(8), 38-40.

Toffer, D.E. (1998). "Building a better pipeline with GIS." *GIS World*, 11(10), 52-54.

Webb, R. (1992). "Networking a regional utilities information system." *Networking Spatial Information Systems (Revised Edition)*, Belhaven Press, West Sussex, England, 171-182.

Wilson, J.D. (1998). "Enterprise resource planning creates new opportunities for GIS in utilities." *GIS World*, 11(5), 51-53.

CHAPTER 9

Stormwater and Waste Management

M.T. Herzog and John W. Labadie

Handling a community's waste streams can be one of the greatest challenges faced by urban planners and engineers. Many urban areas are finding geographic information systems (GIS) to be an important tool to reduce costs of installation, expansion, and maintenance of stormwater and waste management systems. This chapter is divided into two sections. The first section focuses on methods of GIS utilization in stormwater and wastewater system administration and includes applications for maintenance, hydrologic modeling, flood control, and stormwater quality. The second section considers GIS applications in solid and hazardous waste management. GIS uses for managing a landfill from siting through closure is the primary focus, since a portion of solid and hazardous wastes from almost all communities ends up in a landfill. For hazardous waste, meeting RCRA, CERCLA, and NEPA regulations is also addressed, along with a new emphasis on environmental legislation.

INTRODUCTION

Geographic information systems (GIS) have become indispensable in addressing complex environmental issues. Chiefly because "complex spatial operations are possible with GIS that would be very difficult, time consuming, or impractical otherwise" (ESRI 1991). The spatial nature of many stormwater and waste management problems makes them particularly amenable to solution by GIS techniques. Furthermore, waste management has become a controversial topic, due to prevailing NIMBY (*Not-In-My-Back-Yard*) sentiments. Therefore, GIS is gaining use in siting landfills and hazardous waste facilities by providing an efficient means to incorporate community input and feedback in decision-making.

In stormwater and waste mangement, the full power of GIS is often best realized in conjunction with application-specific models, customized graphical user interfaces, external databases, and other computer-based problem-solving methods. Such integration of GIS may be called a *spatial decision support system (SDSS)*. By possessing an entire toolbox of integrated problem-solving elements, including GIS, better solutions can be developed with potential for gaining more rapid public acceptance. Explanation of GIS technology (spatial data sources, spatial data structures, georeferencing systems, operations, integration, and other important aspects) are found in Chapter 2. In the current chapter, each GIS application is shown as a basis for further GIS developments with increasing depth and breadth. This illustrates how GIS can grow and adapt to the increasing complexity of modern urban management problems.

STORMWATER/WASTEWATER APPLICATIONS

AM/FM/GIS

GIS can be an important component of a comprehensive automated mapping and facilities management system (AM/FM). In contrast to alternative AM/FM bases, GIS can efficiently analyze relationships between layers or develop classifications of data into related zones. GIS spatial modeling results can be almost immediately visualized to aid in timely management decisions. In Redlands, CA, a master plan for water and wastewater utilities was developed as a joint effort between the contractors for each utility and the Environmental Systems Research Institute (ESRI) -- a large GIS software and services provider headquartered in the city (ESRI 1997). Customer and parcel level demand data were linked in the GIS to pipe, pressure zone, sewershed information, and other data layers to allow utility integration and analysis.

As another example, the South Australian Water Corporation (SAWC) found that the speed of queries tripled and functionality improved significantly after implementing a GIS defining regional water and wastewater network features. The SAWC is also employing GIS spatial modeling routines for decision support, resource and network analysis, and digital terrain modeling (Carlson-Jones 1996).

The ability of GIS to perform spatial queries can be especially useful when it addresses the integrated mapping requirements of all municipal departments. Consider the growing popularity of underground detention systems as an alternative to surface detention storage in ultra-urban areas (Roberts 1997). A GIS could be used to determine areas where an underground system is needed most for stormwater peak attenuation and/or water quality control purposes. Then the city-wide GIS could be further employed to calculate and display least-cost locations and orientations for design based on soils, electrical, transportation, stormwater, and land-use data layers. Coordination between departments is also enhanced through AM/FM/GIS. For example, sewer system pipe replacement could be scheduled to compliment street repair, with such information exchange enabled through a common GIS basis.

Urban managers preparing to launch into applications of AM/FM/GIS need to look beyond the short-term problems it can address and focus on its long-term utility. Even though GIS software and hardware are now affordable and easy-to-use, there are always data conversion and collection costs during the start-up phase that may be significant. It should be kept in mind, however, that accurate base maps and careful systems development can serve far into the future. For this purpose, there are a growing number of readily available spatial data resources, but most spatial data from outside sources still needs to be updated or expanded to meet municipal AM/FM/GIS needs.

To accomplish this, the Miami-Dade Water and Sewer Department dispatches GPS crews with scanned images on pen-based maps to locate structures and features for developing network models in GIS (Klimas 1997). The city of Portsmouth, VA, imports airborne digital imagery and other remote sensing data, including near-infrared color bands, into a GIS to better estimate impervious area for hydrologic modeling (Phipps 1996a). Other cities have updated U.S. Census Bureau TIGER files of annotated streets, digitized paper maps, or imported CAD files into GIS for base map development. Once imported, graphic or CAD features are linked to a feature attribute table manually by one-to-one association of each feature with a record in the table, or automatically, if CAD text associated with each feature was captured that directly corresponds to a field in the feature attribute table.

Utility System Maintenance

A maintenance program can be organized and efficiently expedited by linking historical maintenance records to a GIS. A maintenance crew dispatcher can input a client's address after receiving a complaint that queries the GIS to pinpoint the problem site and associated pipes for investigation (Phipps 1996b). Database records linked to the map indicate past problems of the associated sewer segment(s) or its surroundings to aid in scheduling equipment and personnel for the type of investigation required. Problem sewer segments can be noted by complaint frequency or service requirements recorded over time and scheduled for investigation and analysis prior to occurrence of an emergency situation.

Milwaukee's Department of Public Works employs a GIS for mapping and analysis in their Work Information Tracking System (Crownover 1991). GIS assists in complaint registration, crew dispatching, scheduling, and geographic analysis. Cass Works (RJN Group, Wheaton, IL), an integrated infrastructure management software package, interconnects GIS with a database management system, allowing problem areas to be plotted on orthophotos for use in the field (Shamsi 1996b). Later, results of the investigation, such as video clips of televised sewer sections, can be launched by simply selecting the sewer segment involved on the GIS display.

Reducing Costs and Improving Efficiency

GIS can be applied in a variety of ways to reduce overall system costs and increase efficiency. The Metropolitan Sewer District of Greater Cincinnati, OH, had a serious infiltration and inflow problem. The problem was largely due to unauthorized connections of roof downspouts and area drains, causing drain overflows and flooded basements (Sweeney 1997). A GIS is used to delineate study areas in which surveys, investigations, and data collection activities are conducted, including smoke and dye testing, sewer televising, and flow monitoring. On each field inspection, sewer locations, conditions, and flows are determined, followed by any necessary maintenance and checking of billing data. This information is then entered into a sewer system attribute database and integrated with the GIS for tracking job order progress. The GIS also spatially integrates and analyzes the available data to correct errors and improve quality. This overall process led to the discovery and correction of over 19,000 unauthorized connections in conjunction with a property owner reimbursement program.

Network routing functionality in Arc/Info (ESRI, Redlands, CA) has been used to determine least-cost routes for new pipes (Taher and Labadie 1996). Several factors are used to determine alternative routing costs, including (1) street network coverage defining traffic loading, (2) the existing pipe network with associated pipe

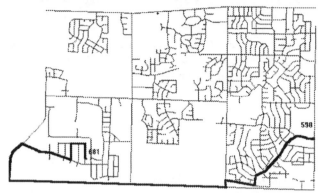

Figure 1 Least-cost routing of pipeline from GIS cost surface (Taher and Labadie 1996)

characteristics, (3) soil types affecting excavation costs, (4) a digital elevation contour map for hydraulic analysis, and (5) land use zoning impacting right-of-way and installation costs. Buffering operations and map overlays are performed to derive coverages with geo-referenced cost information on all pipe routes (Figure 1). In addition to least-cost routing, the GIS employs a network allocation algorithm for assigning service zones to centers based on available pressures and flows. Although applied to design of water distribution systems, this approach could be similarly applied to assign sewersheds to wastewater treatment plants, or to locate and size stormwater detention facilities.

In addition to improving efficiency and decreasing cost, GIS can also improve how utilities are perceived by their clients. Parcels that employ best management practices (BMP) for stormwater control can be rewarded with billing credits to enhance equity in client treatment (Phipps 1996b). To more fairly calculate such billing credits, GIS can be used to help locate all BMP and then more accurately determine their actual impacts on downstream drainage. For example, aerial photos and contour maps can be analyzed in GIS to locate all detention ponds in a community and accurately calculate the size of each.

Systems Integration

Ordinarily, it is not cost-effective for municipalities to integrate GIS with sewer/stormwater models in-house due to the extensive time commitment and programming expertise required. However, there are a number of civil engineering software providers that are dedicated to this effort. The ease-of-use and functionality built into standardized software packages allow periodic software updates that can be rapidly installed on top of existing versions without exacting costly delays. The Massachusetts Water Resources Authority (MWRA) contracted with Harvard Design and Mapping (Harvard MS) to design an automated method for developing geographically correct system schematics for integration with EPA's stormwater management model (SWMM) (Kubaska 1997). The MWRA applies the integrated

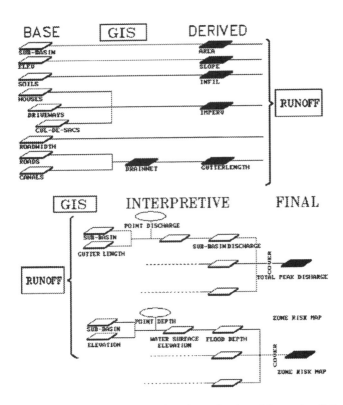

Figure 2   Schematic of GIS layers and analyses developed for each cell for use in hydrologic and hydraulic modeling (Meyer 1993)

SWMM/GIS system to efficiently analyze sewer loading and overflow problems for the city of Boston, and suggest solutions.

In Lexington-Fayette, KY, a GIS for AM/FM operates inside AutoCAD to manage infrastructure inventories (Przybyla 1991). This system is integrated with the HYDRA model for dynamic sewer analysis and the NEWSE model for analysing surcharge conditions that exceed HYDRA's capabilities. The data entered into the GIS serve as model input. The GIS also generates graphical displays and reports of the modeling results. The system has been useful in predicting future growth effects on the sewer system.

Drainage of low-lying areas in Jefferson Parish, Louisiana is controlled by a supervisory control and data acquisition system (SCADA) integrated with a GIS (Thompson 1991). The SCADA monitors dynamic information on water levels and pressures in the drainage, sanitary sewer and potable water networks, while the GIS supplies spatial and static information on the system. The integrated GIS/SCADA system supports a wide range of planning, design, construction, operation, maintenance, monitoring, and data reduction needs of the community. More details on SCADA can be found in Chapter 9.

Urban Hydrologic Modeling

Urban development increases impervious area and alters stormwater flow patterns, which can produce significant increases in downstream peak-flow discharges and stormwater drainage volumes. Prevention of stormwater control system failure requires frequent recalculation of hydrologic response as urbanization extends and intensifies. GIS has become a popular tool in stormwater management for efficiently preprocessing and postprocessing spatially-oriented hydrologic data.

A variety of GIS packages and hydrologic models have been linked, such as GRASS/SWAT (Rosenthal 1996, Srinivasan 1996), GRASS/SMoRMod (Zollweg 1996), GRASS/finite element model & kinematic wave equations (Vieux 1994), Arc/Info/HEC-1 (Robbins 1996), Arc/Info/ANSWERS (Joao 1992), Arc/Info/SWAT (Meherg 1997), Arc/Info/Penn State Runoff Model (Shamsi 1996a), Arc/Info/SWMM (Ceral 1996), Arc/Info/TR20 (ISD 1997), ArcView/SWMM (Shamsi 1996b), IDRISI/RUNOFF (Meyer 1993), and self-developed GIS/PRMS (Battaglin 1996).

In spite of the variety of GIS/model combinations, all of them use a similar procedure for hydrologic modeling. Physically based models require, as a minimum, input of percent imperviousness, soil properties, land use, topography, slope, time of concentration, and sub-catchment area. All these watershed attributes are easily calculated and classified within a GIS due to their spatial nature. Some GIS packages contain advanced hydrologic modeling tools to delineate watersheds, calculate slope from elevation maps, measure flow path length, and fill small depressions that would otherwise confuse results. Raster-based conversion (i.e. from a vector system of polygons into a grid of uniform cells) of all geospatial data layers is usually preferred. This permits direct model input of unique indices (e.g. runoff curve numbers) which are calculated in the GIS by combining information from several spatial data layers with non-spatial parameters (Figure 2).

Comprehensive analysis of inputs, including land use estimates, topography, soils, and grid cell size chosen for the raster data model reveals that data collection methods and data quality may have a profound impact on runoff volume, peak runoff, and soil erosion calculations (Engel 1996). Automating data input processing through GIS improves the consistency of results over traditional methods for watershed delineation, slope, area, and time of concentration calculations (Robbins 1996). Woolpert (Charlotte, NC) uses scattergrams of lag time vs. acreage and a rule-based error-checking expert system within GIS to further improve quality of results.

Although raster cells greatly simplify data layer integration, they cannot accurately depict angular features without resorting to small cell sizes, which results in excessive computer storage and processing requirements. However, by associating each raster cell with an attribute defining the feature type it actually represents, raster topology can better reflect reality (Meyer 1993). The most common raster cell format is single-value attributed data (SVAD), such as assigning homogeneous land use information (i.e. character, string, or numeric value) to each cell. A cell on the edge of a roof top or a parking lot can be assigned edge probability attributed data (EPAD) representing the percentage of impervious area in that cell (Figure 2). Another useful technique is defining a tabular attribute of a raster cell to be a central attributed raster line (CARL) to correctly define features that do not match the raster cell dimension.

Urban Flood Control

Once a hydrologic model has been satisfactorily developed, it is used to determine peak discharges at the outlet of each subbasin for use in flood routing (Figure 2). Hydraulic modeling is required to predict floodwater stages, usually with a probabilistic recurrence interval ranging from 2 to 500 years. Defining and managing flood plains is critical in urban areas because of the high costs associated with flooding residential and commercial areas and the high risk of fatalities. A detailed delineation of at least the 100-year flood plain is also required for the Federal Emergency Management Agency (FEMA) flood insurance study (FIS) program used to determine flood insurance premiums. As with hydrologic modeling, GIS is also employed in hydraulic modeling for preparing input data files and displaying modeling results, such as extent-of-flooding.

The water surface profile models most commonly integrated with GIS are HEC-2 and HEC-RAS developed by the Army Corps of Engineers and SWMM developed by the EPA. Spatial input data required for hydraulic modeling include detailed information on hydraulic structures, channel geometry, and estimates of flow resistance, usually expressed as Manning's n roughness coefficient values. A sewer system inventory yields the required structures information, while field surveys of each reach can provide acceptable Manning's n values.

Developing several miles of channel cross sections by surveying or manual calculation techniques can be costly and time-consuming. Some FEMA FIS contractors use BOSS RMS, an AutoCAD-based River Modeling System developed by Boss International (Madison, WI), to automate HEC-2 input and output in a GIS fashion. Desired cross section locations are indicated on a contour map in AutoCAD, which automatically generates cross section details in a separate viewing window. Non-spatial data are entered in forms associated with each cross sectional view. The RMS menu system generates HEC-2 data directly from the graphical and tabular inputs and executes the model in the background. Profile plots and reports can be generated and flood extent and depth automatically can be interpolated and drawn on the original topographic map display. In a FEMA FIS study completed by Foothill Engineering Consultants (Golden, CO) for the city of Casper, WY, the city's roads and houses were imported using ESRI ArcCad (ESRI, Redlands, CA) and superimposed on the 100-year flood plain delineated with BOSS RMS. This revealed the locations where buildings and roads may be impacted by flooding (Figure 3).

A number of companies have developed integrated model/automated input and mapping packages with similar capabilities to BOSS RMS. KBN Engineering (Gainesville, FL) integrated GIS and SWMM to compare present and future build-out flood maps for Volusia County, FL (Ceral 1997). Woolpert (Charlotte, NC) estimated that integrating GIS and HEC-2 reduced time and cost by up to 25 percent by streamlining tasks and improving quality assurance/control mechanisms (Robbins 1996). Displayed maps indicate areas most threatened by flooding if urban development is permitted, thereby providing the basis for a comprehensive basin master planning.

Spatial data queries can identify promising *what if* scenarios for planning and prioritizing flood management improvements (Phipps 1996a). A GIS was used to map depth-of-flooding results from a HEC-2 analysis for the Trinity River Project of North Central Texas, including the Dallas/Fort Worth area (Promise 1996). The depth-of-flooding map is combined with data on 76,000 structures categorized by jurisdiction and land use to determine the impacts from various flood scenarios, that assist in benefit-cost analyses of flood mitigation proposals. The building structure database generated with GIS not only serves the existing project, but also assists facilities planning and provides permanent benchmarks for commercial surveyors and developers.

GIS can be an important component of a flood warning system to mitigate damages when major flooding is imminent. During a flood in January of 1995 in the Province of Gelderland, Netherlands, GIS was used to generate maps of forecasted flood extent and impacts (Akkers 1995). The required evacuation rates and routes

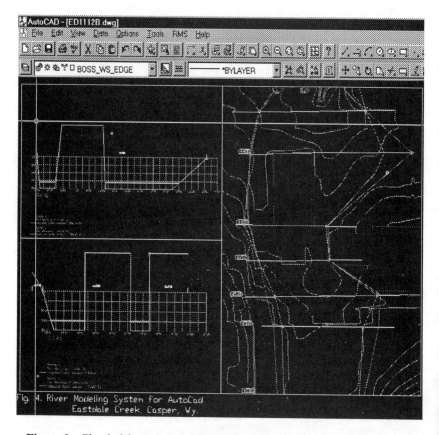

Figure 3   Flood plain analysis for Eastdale Creek, City of Casper, WY (Courtesy of Foothill Engineering Consultants)

158

were rapidly determined with the GIS, as well as dike locations likely to require immediate reinforcement. It is believed that the GIS allowed critical decisions to be made with sufficient lead-time to avert disaster.

After occurrence of a major flood, an opportunity exists to better calibrate models and plan future mitigation efforts more effectively. Flooding on the Missouri and Mississippi Rivers in the summer of 1993 was devastating to many communities. GIS is being used to visualize how the flood developed in these major river basins from January through September of 1993 (White 1996). Geographic animation allows the temporal aspect of the flood development to be better understood. Each of the 273 maps is viewed in sequence to visualize how water leaving one gaging station contributes to flooding of downstream areas. The GIS also displays velocity changes on the maps by adjusting the speed of movement of a symbol for each reach.

Stormwater Quality Monitoring and Control

In 1987, amendments to the Clean Water Act were instituted that require states to assess nonpoint source pollution problems and develop management plans for their remediation. As a follow-up to this Act, the National Pollution Discharge Elimination System (NPDES) was put in effect in 1990. The NPDES requires municipalities with populations over 100,000, select industries, and wastewater treatment plants to obtain permits for discharging wastes into natural waterways. This permitting process can be costly and time-consuming. A city must describe its storm drain system, characterize the water quality throughout the system, and provide the quantity being discharged. Urban areas under 100,000 habitants may have to complete less intensive monitoring if further legislation is adopted, many smaller communities have already begun to develop their own stormwater master plan and guidelines (Veal 1996).

GIS can reduce the costs of meeting NPDES requirements. The hydrologic and hydraulic model links described previously can be used to develop discharge quantity estimates needed to satisfy permitting requirements. GIS can also assist in determining locations and quantities of pollutant accumulations. GIS has been used to estimate watershed pollutant loadings to the Santa Monica Bay, CA (Wong 1997). Although 43,015 water quality data points were recorded for the watershed, the measures were predominantly based on unreliable *grab samples*.

As a more reliable alternative, the EPA database of composite samples taken from known land uses was correlated with land use information derived from aerial photography within the GIS. Watershed pollutant loadings were calculated by overlaying land use, subbasin, and catchment coverages, and relating the results to rainfall and runoff attribute data. The results of the GIS analysis were then used to evaluate, and compare alternative BMP and arrive at the best ones for reducing pollution in urban runoff discharged into Santa Monica Bay. In Huntington, WV, GIS was used to reduce the monitoring requirements for the city's 23 combined sewer overflow discharge points (Shamsi 1996b). GIS was used to select just six of the original sites (i.e. only 25%) to provide a uniform distribution of monitoring equipment throughout the sewershed for accurate runoff water quality modeling.

In the Mill Creek Watershed of Ohio, combined and sanitary sewer overflows, industrial activity, leachate from 30 landfills, wastewater treatment plant effluent and other sources have contributed to poor water quality during wet weather overflows (Sweeney 1997). GIS is being used in a watershed demonstration study funded by an EPA grant. The grant will assist the 25 watershed municipalities to integrate and reference their data and begin to plan collaborative pollution abatement efforts. Phosphorous sources from six towns near the Lakes Region in Maine have been mapped in a GIS (Breau 1997). The Lakes Environmental Association has developed a *Hotspots* model that uses proximity to water bodies, soil types, and slope calculations to help determine phosphorus export coefficients. Phosphorus is a major cause of eutrophication in lakes, so determining how much phosphorus is being transported can aid in lake preservation planning. The program operates under an EPA nonpoint source pollution grant, and GIS is currently being used to develop a focus for phosphorus pollution mitigation.

Intensive monitoring efforts have been conducted at Florida International University using state-of-the-art equipment to monitor individual storm events (Tsihrintzis 1996). GIS organizes the spatial information and provides timely parameter estimates. The high quality data is used to calibrate and verify water quality modeling results. The study results should be useful in characterizing pollutant concentrations in larger stormwater catchments. Another monitoring program involving the entire North Central Texas region includes 30 automated water quality monitors for assessing urban runoff impacts through linkage with GIS-based spatial data layers that include 5,000 major outfalls (Promise 1996).

INTEGRATED SOLID WASTE MANAGEMENT

Overview

Solid waste management is "the discipline associated with the control of generation, storage, collection, transfer and transport, processing, and disposal of solid wastes in a manner that is in accord with the best principles of public health, economics, engineering, conservation, aesthetics, and other environmental considerations, and that is also responsive to public attitudes." Integrated solid waste management (ISWM) is "the selection and application of suitable techniques, technologies, and management programs to achieve specific waste management objectives and goals" (Tchobanoglous 1993).

The engineer or planner is faced with an imposing task to find methodologies that can effectively address all the required components of an ISWM program. GIS may be the answer to this dilemma since it can provide assistance at every stage of ISWM. The first step in managing solid wastes is developing an understanding of city/county-wide patterns of waste generation and handling activities. In the past, this may have been accomplished by random surveys, with results entered into a database. Unfortunately, the reports and charts generated from these surveys tend to aggregate results and therefore obscure spatial patterns in the generation, storage, and transport questions to be addressed. With the spatial database incorporated in a GIS, however, the amounts and types of wastes can be assessed for every sector and neighborhood, and visually analyzed for planning and management purposes.

For example, in some communities an excessive number of private trash haulers may share similar routes, thereby disturbing residential neighborhoods and causing excessive street wear. GIS can be applied to design plans for awarding exclusive residential trash hauling contracts for specified geographic districts that would maintain average trash volume and profits for current haulers. GIS allocation routines can assign each hauler to the correct areas required to fill the trucks. GIS could also be used determine where to locate a waste transfer station closest to the center of mass of waste generation for satellite communities removed from the urban center.

Street network routing algorithms are included in many GIS software packages that can be used by municipalities and private trash haulers to develop the most efficient collection routes. Stoplights, left turns, cul-de-sacs, distance, and other route attributes that increase collection time or reduce efficiency (e.g. hauling partially filled containers) are assigned cost factors. The network algorithm minimizes the overall total of the cost factors to ensure that a hauler follows the least cost path. RouteSmart (Bowne Distinct 1997) employs network algorithms in a popular commercial GIS along with a variety of geographic information to optimally route sanitation vehicles. The municipalities of Philadelphia, PA, applied RouteSmart to reduce its fleet size by 22%, leading to a sizable reduction in vehicle use, staff, and operating expenses.

One of the most critical needs that GIS can serve in ISWM is siting and managing a landfill. Although there are a number of waste-to-energy projects deployed throughout the U.S., most waste that is not recycled or reused is still conveyed to a landfill. A landfill is a technologically advanced system that prevents the migration of leachate and gases generated from decaying wastes, inhibits vermin and insect problems, and allows the land upon which it is built to be restored to productive use after closure (such as a park). With increasing land use pressure and sensitivity to landfill impacts, siting new landfills has become a complex and time-consuming undertaking. Often, a task force that includes environmentalists, developers, trash haulers, municipal and county planners, and citizen groups is organized to participate in the siting issue. Concurrence among such a diverse committee is only possible if all their needs receive consideration in selecting a site, in addition to the array of legislated mandates. Otherwise, time and money can be wasted on investigating sites that later turn out to be unacceptable to key groups in the approval process.

GIS can efficiently accomplish the daunting task of landfill siting through a phased approach. In the first phase, screening criteria that eliminate a site as unacceptable are overlain to reduce the size of the area for which additional data will need to be considered (Lindquist 1991). Such exclusionary data may include legal restrictions (e.g. proximity to airports - a flocking bird danger), physical impracticability (e.g. on a mountain top or in a lake), threats to public water supply, proximity to residential areas, location in flood plains or wetland areas; overlying geological fault zones, or impacts on threatened species or plants.

In the second phase, GIS can be used to rank and normalize nonexclusionary data of concern to the community (Siddiqui 1996). Non-exclusionary criteria usually include site size, compactness, soil effectiveness for cover and foundation, land use and cost, average hauling distance and time, slope, elevation, secondary environmental impacts, surrounding developmental pressures, and other site engineering factors. Normalizing the data in a GIS can be accomplished by calculating the relative weight of the value of a criterion at any location as a relative percentage of the range that the criterion can achieve. This can be multiplied by a relative importance value of one criterion in relation to all the others. A simple weighted ranking can then be determined by summing all the layers and ordering the site alternatives from the most favorable to the least favorable alternative (Kao 1996a).

Since this method can be highly subjective, other engineers have chosen to extract the data layers defining criteria for each unit area in the form of a matrix that can be used in a variety of multi-criteria decision analysis (MCDA) algorithms. MCDA may better ensure that all criteria contribute proportionally to their import in the final site selection to maximize objectivity.

An example of the GIS/MCDA method was presented for Cleveland County, OK (Siddiqui 1996). The MCDA chosen to optimize the site selection is the analytical hierarchy method (AHP) (Saaty 1980) which is recommended by EPA for comparing decisions (EPA 1993). AHP attempts to decompose the problem into manageable subproblems linked in a decision tree or *hierarchy*. Modern AHP programs efficiently carry out the sorting, ranking, and aggregation of measures that lead to a final suitability index, which provides an absolute rank of the overall desirability of each site. The method includes a formula to check the decision-makers consistency in comparing the desirability of alternatives. A pilot landfill siting study for Larimer County, CO, utilized two other popular MCDA techniques: Promethee (Preference Ranking Organization METHod for Enrichment Evaluation) and compromise programming (Herzog 1997). By employing two established MCDA methods, the robustness of the analysis results can be further tested.

Promethee is based on comparison of the positive and negative preferences of one alternative over another for each criterion. Compromise programming is a popular method because of its simplicity and its capability to evaluate a set of alternatives from several different perspectives (Tecle 1992). The siting analysis process for Larimer County was developed into an SDSS that includes a graphical user interface within the GIS from which the MCDA algorithms are executed. The SDSS allows the user to interactively adjust parameters such as landfill life and projected daily solid waste volume (Figure 4). The results are presented in the form of summaries, output tables, or GIS map displays of the selected site. Through community feedback and MCDA integration, GIS ensures that each interest is weighted fairly in final siting decisions. As an alternative to the use of MCDA methods in the ranking phase, an expert system can be used to enhance GIS analysis by incorporating criteria into a rule-based structure for site elimination and prioritization (Kao 1996b).

Landfill Design, Operation, and Maintenance

The GIS developed for site selection can be enhanced to assist in design, operation, and maintenance throughout the life of the landfill. All criteria layers used to site the landfill can be subsequently applied to management decisions. For example, the elevation layer developed during the siting phase can be used to derive aspect *coverages* for conducting visual impact analyses. These may be used for planning the height of berms and vegetative fences required to screen the landfill from the view of surrounding roads and habitations.

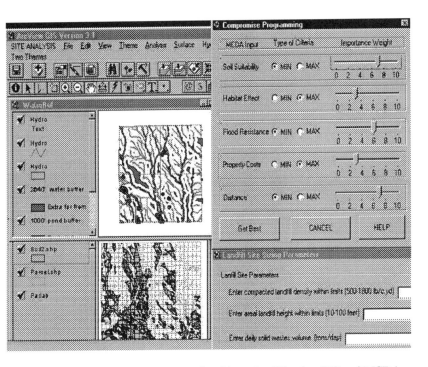

Figure 4 Decision-support system for siting a landfill using GIS and MCDA

GPS-based topographic and features data can be directly input into GIS to assist in drainage and vehicle routing, as well as expansion and closure activities for landfills. GIS can also serve in landfill design and management by (Green 1996):

- calculating the area of geomembrane liners
- estimating the volume of clay needed for lining
- locating gas and leachate pipes
- determining required pipe lengths
- recording waste composition and compaction to optimize density
- evaluating the potential for gas migration
- analysing the chemical and physical stability of the site over time
- recording and analysing the filling process to maximize capacity while minimizing environmental impacts.

In New Jersey, besides all the landfill design and operations requirements of the Federal Resource Conservation and Recovery Act (Subtitle D), the state environmental department requires an annual topographic update with volume certification for all landfills (Fralinger 1997). Albert A. Fralinger, Jr. Consultants (Bridgeton, NJ) found that differential GPS could be used to survey the Salem County, NJ, Landfill accurately and in an efficient, cost-effective manner. The

163

estimated 0.79-inch accuracy of the GPS results exceeds that of aerial photography, without requiring a total station operator to remain in the landfill for several hours/days. The county saved approximately $150,000 through this procedure by confirming that construction of a new cell could be delayed for another year.

Landfill Closure and Post-Closure

If GIS has been used in landfill management from landfill inception, it becomes a valuable tool in the closure and post-closure phases. It can aid in planning and managing development of the soil/synthetic membrane cap, drainage grid and stormwater containment system, and revegetation. It can be used to determine the best locations for required groundwater contamination monitoring wells and post-closure settlement monitors. It can also be used to predict where water will collect and pond before the added weight increases differential settlement or water migrates into the wastes, increasing leachate production. If density data have been developed in the GIS throughout the life of the landfill, settlement can be more easily predicted, which is crucial for planning the ultimate use of the reclaimed land.

It is best to dedicate closed landfills to *soft* uses such as parks, agricultural areas, golf courses, natural areas, and light structures. Although GIS can be invaluable in selecting and designing these uses, GIS records become even more important if development pressures force the land to be zoned for *hard* uses (e.g. heavy structures or paving). In this case, GIS can help determine gas extraction requirements, methods to combat settlement, and ways to assuage environmental concerns.

Intensive post-closure monitoring upgradient and downgradient of a landfill to detect leachate migration is required for many years after landfill closure in the U.S. The reason is that leachate from municipal wastes may contain a variety of hazardous chemicals (Montague 1988). Fortunately, many modern landfills are designed with two or even three essentially impermeable synthetic liners augmented by intermediate layers of low-permeable clay and drainage media. When properly installed, these systems completely encase the waste and intercept and remove *run-on* waters that have migrated to the site. With little or no water entering such a system, almost no leachate is produced, and the wastes may actually desiccate. Most double-liner systems also include a detection system designed to evaluate the quality of water exiting the top liner. Studies have shown that leakage from such modern double-liner systems is minimal (Reddi 1996). GIS may be used to analyze and present such double liner study results to convince regulators to reduce the unnecessary monitoring burden in such situations. The savings could be applied to reduce landfill tipping fees or taxes, or pay for attractive end-use development that better serves the community.

HAZARDOUS WASTE MANAGEMENT

Resource Conservation and Recovery Act

The use of GIS in hazardous waste management is predominantly associated with compliance and/or remediation issues. Three major congressional acts have established the foundation for these efforts. The Resource Conservation and Recovery Act (RCRA) of 1976 created the manifest system for documenting the handling of all hazardous wastes in the U.S. from production through ultimate disposal (*cradle-to-grave*). Producers, transporters, and disposers of hazardous waste are held responsible for proper management of the wastes based on legal guidelines and a permitting system. GIS is used to organize industrial, military, and public programs to satisfy the RCRA requirements. For example, many laws have been promulgated to govern hazardous waste transport from generation to disposal site. GIS has been used to select the best routes that minimize exposure risk based on these federal requirements (Tsihrintzis 1996).

GIS can also aid in the establishment of proper facilities to handle hazardous waste. GIS was used to site a low-level radioactive waste facility for the four state region of Pennsylvania, Maryland, West Virginia, and Delaware (Dwyer 1996). Law Engineering (Harrisburg, VA), with responsibilities for site selection, developed 35 exclusionary criteria layers in a GIS for the entire state of Pennsylvania similar to smaller scale landfill site selection methods described previously. Data preparation represented a major task, requiring 60 digitizers, GIS batch programs for efficient map manipulation, and a method to remove each area from further consideration as soon as only one criterion disqualified it. The GIS was used to produce over 10,000 maps for public agencies, allowing prompt dissemination of information to the public as certain areas of concern were eliminated. This process, along with extensive public involvement during the criteria development stage, bolstered public trust.

CERCLA Act

The Comprehensive Environmental Response, Compensation, and Liability Act (CERCLA) of 1980, or *Superfund*, was enacted to evaluate and clean up hazardous contaminated sites abandoned since the beginning of the industrial revolution. It requires contaminators to pay for hazardous site remediation years after the contamination occurred, albeit often unwittingly. GIS is well suited to organize the large volumes of environmental and site data required for site investigations and remediation efforts. GIS allows spatial relationships between soil, water, and air environments and wastes to be explored and modeled. GIS is also used to review site history long after the damage has occurred in order to reconstruct hazardous waste disposal scenarios and effectively plan monitoring and clean-up efforts. GIS provides an efficient means of updating public and legal concerns on Superfund project progress, reducing costs associated with litigation and public resistance, thus expediting the clean-up process.

ERM Program Management (Exton, PA) has found GIS to be an important tool in conducting site investigations (Douglas 1996). Base maps are constructed in the GIS with information layers on soils, geologic stratigraphy, sediments, water, chemical analysis, topography, groundwater elevations, and site features. GIS assists in selecting well field locations and posting groundwater elevations to create an interpolated map of groundwater elevations. GIS also facilitates extraction of data for review, analysis, and comparison with compliance laws, while maintaining coordination among the many disciplines involved.

Although 40,000 Superfund sites were originally included in an inventory of potentially hazardous sites, over 27,000 sites have recently been removed from the list for being *insufficiently contaminated* (Maldonado 1996). GIS is being applied to analysze these unused industrial sites of uncertain hazard, or *brownfields*, in a number of local and state programs designed to prioritize them for possible redevelopment. In Bridgeport, CN, GIS also serves as a potent marketing tool by allowing potential brownfield developers to query for specific property qualities.

Somewhat sheltered from financial problems and litigation, the U.S. military is at the forefront of contaminated site remediation. An example in which the military has effectively used GIS to cleanup polluted military bases, as uncovered during recent closure activities, is the Griffiss Air Force Base, NY (ESRI 1995). Old aerial photos and archived information were used to reconstruct industrial site locations now hidden by development and overgrowth. The Arc/Info GIS package was utilized by LAW Environmental (Atlanta, GA) to create a menu-driven interface for chemical database queries, document management, cartographic production, and spatial analysis. Many different sources of information can be cross-correlated using the system to aid in critical environmental impact mitigation decisions. At some bases, leaking underground storage tanks are of primary concern, such as at the Fort Devens, MA, site. Here too, the primary use of GIS is to reconstruct past operational patterns at the Fort to construct the best schemes for future clean-up efforts. The hard copy maps and digital files developed with GIS also assist clients with other types of redevelopment tasks.

The Army Corps of Engineers has combined GIS with neural networks and a knowledge-based system (Millhouse 1997). This SDSS can be applied to locate buried munitions, including unexploded artillery shells, and assist in organizing removal efforts. GIS is used to evaluate information on more than 2,000 sites inventoried in this ordnance clean-up program. The GIS also manages the ordnance and explosives program for each of the 60 to 80 active projects being explored at any one time, determines which sites require further investigation, sorts out ordnance from background anomalies present in remotely-sensed data, manages removal, and organizes records. It also provides the framework for the analysis of the complex neural network and fuzzy logic programs involved in discriminating ordnance from background effects. This effort reduces the need for intrusive, destructive, and costly excavation operations and improves safety, while simultaneously enhancing the effectiveness of munitions recovery efforts.

National Environmental Policy Act

The National Environmental Policy Act (NEPA) of 1969 requires all projects that receive a portion of federal funding to develop environmental impact statements (EIS). EIS serve to educate and inform the public on how the environmental effects of a project have been considered in its development. GIS can be used to help organize the data collected from the multiple disciplines usually involved in EIS development. An impact assessment system to create federally compliant environmental impact information was developed using tables integrated with GIS through a graphical user interface (Youngs 1996). Fifteen databases with a user-friendly *front-end* are used to collect different types of data that could have environmental implications. The system also stores historical records for estimating cumulative environmental impacts over time.

New Emphasis on Environmental Legislation

It appears that the so-called *regulate/enforce* approach that led the U.S. EPA to enact RCRA, CERCLA, and NEPA legislation has fallen into some disfavor. There is indication that this regulatory emphasis may be replaced by a *plan/audit* approach that focuses remediation on land use and encourages application of more innovative technologies (Dunbar 1996). In the new legal environment, GIS will continue to play an increasingly prominent role in hazardous waste site characterization and clean up.

Another trend in environmental protection on all levels is a focus on community participation and education. GIS can assist in viewing the environment in an integrated way, and help citizens play a more active role in enhancing the quality of air, water, and living resources while maximizing resources in a way that spurs economic development (Kapuscinski 1996). The EPA has taken a proactive role in this regard by establishing a website for the National Geospatial Data Clearinghouse (http://nsdi.epa.gov/index/html) and its own Envirofacts Warehouse at http://www.epa.gov/enviro/html/ef_home.html. These sites include superfund and hazardous waste data, the toxic release inventory, GIS-based maps, online query forms, and a host of educational and useful environmental information for anyone to peruse or download.

CONCLUSIONS

This chapter should have revealed GIS to be a powerful tool for managing stormwater and wastes, but as any application GIS was shown to be most effectively utilized by integrating it with other important modeling and management software. Just as GIS itself is a combination of database, mapping, analysis, and visualization components, we can leverage its utility through the effective methods that exist today for making disparate software function in a single problem-solving framework. GIS utility is further accentuated today by extending the user base through more intuitive easily learned user interfaces, greater access through the Internet/Intranets, and a growing variety of other information-sharing methods. GIS itself appears to be evolving at an increasing rate as the number of application providers swells. This can only ensure that the applications discussed in this chapter will be quickly over-shadowed by more functional applications that will be even easier to use in spite of their ever-expanding capabilities.

REFERENCES

Akkers, B., Overduin, T., Overtnars, B., and Voet, P. (1995). "Escaping the water." *ArcNews*, ESRI, 17(3), 28.

Battaglin, W., Kuhn, G., and Parker, R. (1996). "Using GIS to link digital spatial data and the precipitation runoff modeling system: Gunnison River Basin, CO." *GIS and environmental modeling*, GIS World Books, Fort Collins, CO.

Berry, J. (1993). *Beyond mapping - concepts, algorithms and issues in GIS.* GIS World Books, Fort Collins, CO.

Bowne Distinct, LTD. (1995). "RouteSmart is helping to clean-up Philadelphia." *ArcNews*, ESRI, 17(4), p 36.

Breau, S. "Using GIS to identify phosphorus threat." *ArcNews*, 18(1). p. 38.

Carlson-Jones. J, (1996). "The future is clearer for SA water drinkers." URL: http://www.esri.com.

Ceral, T., Tremwel, T. and Bureson, R. (1996). "Use of Arc/Info, EPA-SWMM, and UNIX text processing tools to determine flood extent." *1996 AWRA symposium on GIS and water resources*, AWRA, Madison, WI.

Crownover, L., and Lior, S. (1991). "Milwaukee's work tracking IMS: building upon a geographic foundation." *Civil engineering applications of remote sensing and geographical information systems*, ASCE, NY.

Douglas, W. and Maitan, I. (1996). "Accessible information." *Civil Eng.*, ASCE, 66(6), 59-62.

Dunbar, G. and Foster, S. (1996). "Strategies for remediation managers." *Civil Eng.*, ASCE, 66(6), 53-55.

Dwyer, J. (1996). "Arc/Info used in disposal facility siting process." *ArcNews*, ESRI, 18(2), 24.

Engel, B. (1996). "Methodologies for development of hydrologic response units based on terrain, land cover, and soils data." *GIS and environmental modeling*, GIS World Books, Fort Collins, CO.

ESRI. (1 997). "Integrating water and wastewater utilities with Arc/Info." *ArcNews*, ESRI, 19(4), 30.

ESRI. (1995). "GIS used to clean up U.S. military's toxic wastes." *ArcNews*, ESRI, 17(2), 12.

ESRI. (1991). *Understanding GIS: The Arc/Info method.* Environmental Systems Research Institute, Redlands, CA.

EPA. (1993). *Life cycle design manual: environmental Requirement and the product system.* U.S. Environmental Protection Agency, Wash., D.C.

Fralinger, C. and Maxwell, J. (1997). "Mapping with a differential." *Civil Eng.*, ASCE, 67(3), 50-52.

Green, D. (1996). "GIS and its use in waste management." *1996 ESRI European User Conference Proceedings*, ESRI, Redlands, CA.

Herzog, M. (1997). "Decision support system for siting a new landfill in Larimer County." *Unpublished study in partial completion for MS program*, Dept. of Civil Eng., Colorado State University, Fort Collins, CO.

Innovative System Developers. (1995). "GEOSTORM: GIS productivity software from ISD." [WWW document] URL http://www.innovsys.com.

Joao, E. and Walsh, S. (1992). "GIS implications for hydrologic modeling." *Comput. Environ. & Urban Sys.*, 16, 43-63.

Kapuscinski, J. (1997). "A community approach to environmental protection." *ArcNews*, ESRI, 19(2), 13.

Kao,J., and Lin, H.(1996a). "Multifactor spatial analysis for landfill siting." *J. Enviro. Eng.*, ASCE, 122(10), 902-908.

Kao. J., Chen, W. and Lin, H. (1996b). "Network expert geographical information system for landfill siting." *J. of Comput. in Civil Eng.*, ASCE,10(4), 307-317.

Klimas, P. (1997). "Field-based GIS/GPS works for water and sewer conversion and applications." *1997 ESRI User Conf. Proceedings*, ESRI, Redlands, CA.

Kubaska, B. (1997). "GIS schematic development and database interface for Greater Boston sewer system model." *1997 ESRI User Conference Proceedings*, ESRI, Redlands, CA.

Levy, K. (1991). "Building GIS relationships spatially and politically." *Civil Engineering Applications of Remote Sensing and Geographical Information Systems*, ASCE, New York, N.Y.

Lindquist, R. (1991). "Illinois cleans up: using GIS for landfill siting." *Geo. Info. Syst.*, Februsry, 30-35.

Maldonado, M. (1996). "Brownfields Boom." *Civil Eng.*, ASCE, 66(5), 36-40.

Meherg, K. (1997). "Framework for identifying critical areas producing nonpoint source pollution in a large watershed using GIS and SWAT". *Proceedings of the 7th Annual Amer. Geophysical Union hydrology Days*, Colorado State University, Fort Collins, CO, April, 212-214.

Meyer, S., Salem, T., and Labadie, J. (1993). "Geographic information systems in urban storm-water management." *J. Water Resour. Plng. and Mgmt.*, ASCE, 119(2), 206-228.

Millhouse, S. and Gifford, M. "Net results." *Civil Eng.*, ASCE, 67(6), 58-60.

Montague, P. (1988). "Leachate from municipal dumps has same toxicity as leachate from hazardous waste dumps." *Rachel's Hazardous Waste News*, 90, 15.

Phipps, S. (1996a). "GIS image technology helps in stromwater planning." *Amer. City & Co.*, 8(6), 48.

Phipps, S. (1996b). "Using raster and vector GIS data for comprehensive storm water management." *AWRA Symposium on GIS and Water Resources*, AWRA, Madison, WI.

Promise, J., Brush, S., Rae, S., and Blanchard, S. (1996). "Improving regional water resources planning with GIS and InterNet: The Dallas/Fort Worth experience." *AWRA Symposium on GIS and Water Resources*, AWRA, Madison, WI.

Przybyla, J. and Kiesler, C. (1991). "Extending GIS capabilities for enhanced sewer system modeling." *Civil Engineering Applications of Remote Sensing and Geographical Information Systems*, ASCE, New York, N.Y.

Reddi, V. and Scarlatos, P. (1996). "Leachate leakage from landfills with modern liner systems." *Water Res. Bulletin*, AWRA, 32(4). 681-696.

Robbins, C., and Phipps, S. (1996). "GIS/water resources tools for performing floodplain management modeling analysis." *AWRA Symposium on GIS and Water Resources*, AWRA, Madison, WI.

Roberts, B. (1997). "Storm-water treatment goes underground." *Civil Eng.*, ASCE, V66(6), 56-57.

Rosenthal, W. and Hoffman, D. (1996). "Hydrologic modeling to aid in locating monitoring wells." *AWRA Symposium on GIS and Water Resources*, AWRA,

Madison, WI.

Saaty, T. (1980). *The analytic hierarchy process*. McGraw-Hill, New York, N.Y.

Shamsi, U. (1995). "Water resources engineering applications of geographic information systems." *Wetland and Environmental Applications of GIS*, Lewis Pub., New York, N.Y.

Shamsi, U. and Fletcher, B. (1996). "ArcView applications in stormwater and wastewater management." *AWRA Symposium on GIS and Water Resources*, AWRA, Madison, WI.

Siddiqui, M., Everett, J, and Vieux, B. (1996). "Landfill siting using geographical information systems: a demonstration." *J. of Enviro. Eng.*, ASCE,122(6), 515-523.

Srinivasan, R., Arnold, J., Rosenthal, W., and Muttiah, R. (1996). "Hydrologic modeling of Gulf Basin using GIS." *GIS and Environmental Modeling*, GIS World Books, Fort Collins, CO.

Sweeney, M., Quinn, T., Quinn, B., Ingram, T., Allen, R., Hammond, R., and Smith, A. (1997). "GIS support of comprehensive water quality improvement programs." *ESRI User Conference Proceedings*, ESRI, Redlands, CA.

Taher, S., and Labadie, J. (1996). "Optimal design of water distribution networks with GIS." *J. Water Resour. Plng. and Mgmt.*, ASCE, 122(4), 301-311.

Tehobanoglous, G., Theisen, H., and Vigil, S. (1993). *Integrated solid waste management – engineering principles and management issues*. McGraw-Hill, New York, N.Y.

Tecle, A. (1992). "Selecting a multicriterion decision making technique for watershed resources management." *Water Resources Bulletin*, AWRA, 28(1),129-140.

Thompson, G., Byrne, P., Hebert., S., and Young, J. (1991). "Integration of SCADA and GIS technologies in Jefferson Parish, Louisiana." *Civil Engineering Applications of Remote Sensing and Geographical Information Systems*, ASCE, New York, N.Y.

Tsihrintzis, V., Hamid, R, and Fuentes, F. (1996). "Use of geographical information systems in water resources: a review." *Water Res. Mgmt.*,10, 251-277.

Veal, S. (1996). "Small cities await flood of stormwater regulations." *Amer. City & Co.*, 8(6), 42-47.

Vieux, B. and Gauer, N. (1994). "Finite-element modeling of stormwater runoff unwing GRASS GIS." *Microcomputers in Civil Engineering*, 9(4), 263-270.

White, W., Mizgalewicz, P., Maidment, D., and Ridd, M. (1996). "GIS modeling and visualization of the water balance duning the 1993 midwest floods." *AWRA Symposium on GIS and Water Resources*, AWRA, Madison, WI.

Wong, K., Strecker, E. and Strenstrom, M. (1997). "A picture worth more than 1,000 words: geographic information system provides fine detail for nonpoint source model." *Water & Environment*, 22, 41-46.

Youngs, D. (1996). Impact assessment to support NEPA regulatory compliance. *Personal communication*, Sept 25, by e-mail: gsri@acca.nmsu.edu.

Zollweg, J., Gburek, W., and Steenhuis, T. (1996). "SmoRMod -- a GIS-integrated rainfall-runoff model." *Transactions of the ASCE*, 39(4), 1299-1307.

CHAPTER 10

Cultural and Natural Resources

Kristina Dalton

Geographic information system (GIS) is becoming the tool of choice for mapping. It has the ability to store large amounts of data, do statistical analysis, and depict the results of the analysis graphically, thus giving it an undeniable edge over two-dimensional (2-D) graphics programs. This chapter presents a few examples of how GIS is being used in cultural and natural resources. Predictive modeling, site risk analysis, three-dimensional (3-D) modeling, line-of-sight and intervisibility are ways in which GIS is being employed in the field of archaeology. Within the natural resources, applications for GIS include vegetation studies, biological evaluations, and mine planning. GIS offers innovative solutions to problems that were previously solved by extensive field work or manual mapping. In some cases, GIS can offer a 3-D interpretation, helping to give a better understanding to complicated issues.

INTRODUCTION

Traditionally, paper maps have been a graphical representation of the physical and cultural environment. For the most part, they were not used for detailed analysis of data, but rather for an initial contact that would give a generalized look at the information. From that point, questions could be asked and data could be studied in detail, independently of the map. Geographic information systems (GIS) have begun to change that method so that spatial data, including the smallest details, can be stored along with its geographic coordinate within the program. Complex analysis can be preformed on the spatial and geographic data and the results can be presented in either tabular or map format.

Uses of GIS in cultural and natural resources are expanding rapidly as the software is becoming more affordable and easier to use. It is now used worldwide as the tool for mapping and solving spatial questions. Throughout the United States, many cities and counties have begun to transfer their paper maps into GIS databases. A good example of this is the city and county of Honolulu in Hawaii, which is in the process of creating and updating coverages that will make up a series of map libraries (Worrall 1990). Not only will it contain information at the parcel level, but it will also contain data on fish catch areas, artificial reefs, marine ecosystems, recreational areas, harbors, navigational routes, bathymetric zones, and waste disposal sites.

Local governments in many countries are bearing more responsibility for effective management of natural resources and are seeking new ways to utilize information. The heaths of Dorest in the United Kingdom have become a focus of a study as pressure from populated areas encroached on the remaining heathlands (Carver 1997).

GIS is helping to evaluate the components that make up the heathlands and to create models to predict future changes. Also, the spatial and temporal distribution of objects is a major topic of study for archaeologists (Ebert and Hitchcock 1980). GIS has also proven to be useful in studying more complicated network patterns. For example, GIS was used to help evaluate the factors which might have determined the locations of prehistoric roads in the American Southwest for the Anasazi communities that existed between AD 900 and AD 1150.

APPLICATIONS

Applications in Cultural` Resources

The field of archaeology lends itself wonderfully to the applications of GIS. A unique artifact is tied to an (x, y) coordinate on the ground. Attributes about the artifact are gathered and entered into a database. As the search widens, groups of clustered artifacts are mapped and site boundaries defined. Eventually a regional picture can be pieced together from compiled information about the sites, landscape, vegetation, climate, and wildlife of the area. All this detail and information can be incorporated into the database and represented as a map. Thousands of pieces of information can be queried, sorted, sifted, and examined. Statistical analysis can be run on the information and then represented in a graphical manner. GIS can do all this and more.

Predictive Modeling

One of the simplest uses for GIS in archaeology can be seen in predictive modeling. Predictive modeling attempts to define areas that are most likely to have archaeological sites. By eliminating areas with poor correlation to known habitation factors, more time can be spent searching areas that have the greatest possibility of containing sites. Themes such as water sources, rainfall, slope, vegetation, geology, food sources, and previously known sites are mapped. Figure 1 shows a simplified example. The information on each theme is ranked by its influence on the habitability of an area. By overlaying the maps and adding the cell values, areas of high correlation are determined. Although predictive modeling can be accomplished using manual methods of mapping, with the use of GIS, it is easier to combine multiple themes, to construct and incorporate slope models, and to add as much or as little detail as needed. The real advantage of GIS is that it allows multiple testing of an area by adding and subtracting layers, relatively easily.

Site Risk Assessment

Along the same lines, another use for GIS is site risk assessment. Utilizing themes such as roads, access, terrain, and location from populated areas an estimation of the possibility of a site being vandalized can be made. Also records can be kept

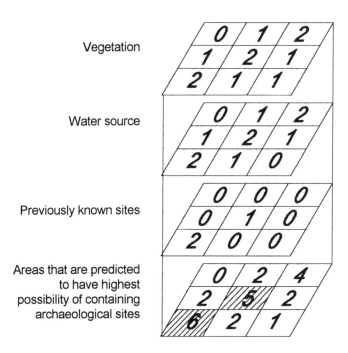

Figure 1 Predictive modeling can be useful in finding areas with a high probability of containing archaeological sites. Using themes that have known habitation factors, such as vegetation, water source, known sites, food source, rainfall, shelter, and slope areas with the highest correlation can be located.

on such factors as erosion, to determine if the site is at risk from the forces of nature.

Since the role of GIS in archaeology is relatively new, we can look forward to seeing some innovations as the technology improves. Currently, the bulk of the GIS work revolves around mapping space/time distributions. For instance, studying settlement patterns in a region, historical ecology, distribution of burial sites, mapping and analyzing, ceramics, stone artifacts, game traps, and studying the response of human groups to vegetation changes. The distribution of these things on the ground and over time is the center of the focus. Each theme can be mapped and then overlain on other themes for further study. GIS is very good at doing this type of study, particularly if there are many themes, or massive amounts of data that needs to be plotted, since plotting ten features or a thousand makes little difference. Recently, GIS has been able to incorporate three-dimensional (3-D) models into its software. This means mapping can now extend from a simple two-dimensional (2-D) piece of paper into a new way of thinking in 3-D.

Line-of-Sight and Intervisibility

One of the new and more complex uses of GIS in archaeology, is a line-of-sight analysis and intervisiblilty. These types of investigations reveal what can be seen from one particular viewpoint or if a network of viewpoints can see each other. Figure 2 shows a seen-unseen analysis that is an example of a type of intervisibility using a close network of points. This is an important aspect for many primitive cultures both in communications with other villages and for defense purposes. For example, a possible site could be eliminated as a signaling position for defensive purposes if it is not in the line of sight of the settlement. Previously, short of going out in the field and hiking the peaks, there was no other way to demonstrate this idea.

Using GIS and 3-D models the idea can be tested in advance and positions can be eliminated or added, thus expediting the fieldwork, and leaving time to concentrate on the most likely areas.

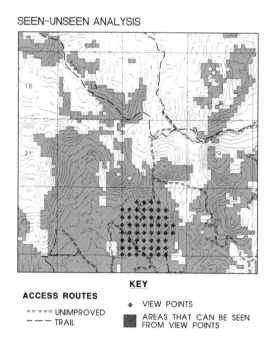

SEEN-UNSEEN ANALYSIS

KEY

ACCESS ROUTES

===== UNIMPROVED
—·— TRAIL

✛ VIEW POINTS

▨ AREAS THAT CAN BE SEEN
FROM VIEW POINTS

Figure 2   Line-of-sight, intervisibility and seen-unseen analysis, provide ways to locate areas on the map that have a visual relationship. By using digital elevation models, they can reveal what can be seen from a viewpoint, or a series of viewpoints.

3-D Modeling

Constructing a 3-D model in a GIS system to test a line-of-sight analysis or intervisibility requires coordinates and elevations for the site location(s) and a DEM (digital elevation model). The DEM is a digital file that contains elevations from which contours can be generated. Many GIS programs can utilize this data to create a model and view sheds from the site locations.

Not only is modeling the terrain possible, but also modeling the cultural features found in the terrain. Figure 3 shows a three-dimensional model of a Hohokan pithouse, found in the American southwest in Arizona circa AD 1100 to AD 1250. This model was constructed from survey information collected in the field. The raw data was converted into a.dxf file format, brought into AutoCAD 12 and converted into a 2-D map. Two more programs were then employed to create 3-D meshes and solid faces, and finally finishing the model by adding lighting, views, and scenes. Short animations including a flyby and walk-through were also created. This is an excellent example of what can be accomplished by using the current technology and a creative archaeologist.

Data Sharing

Data is the limiting factor in any GIS study. For most individuals it would be very expensive, time consuming, and nearly impossible to collect data and create the base maps. A natural progression to the demand for data is leading to data sharing. Recently some states have started inputting archaeological survey information into centralized GIS systems. Deciphering the hundreds of paper maps and associated overlays and processing that information into a GIS system is a long and laborious task. In the future, finding previous surveys and information gathered during studies will be easy to access.

Applications in Natural Resources

Over the last 20 years the term natural resource has changed to encompass a larger view. Previously it referred to the resources provided by nature and considered an economic benefit to a country, such as mining, fishing, logging, agriculture, and energy sources. The current focus is tending toward the more intangible assets, such as endangered plant and animal species, wetlands, wildlife management, parks, recreation, and clean air. These types of resources are more difficult to define, measure, and map. And yet, the challenge remains to do so.

Vegetation Studies

Vegetation studies have become popular in recent years, especially studying vegetation changes through time. Typically, a study would be setup by acquiring aerial photography and developing a classification system for the vegetation. Figure

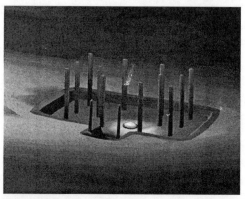

Figure 3   Example of modeling a cultural feature is shown here as a 3-D model of a Hohokam pithouse. It was created from a 2-D field map. Currently this is a time consuming process, but much can be learned from the results.

Vegetation Mapping

Aerial Photo

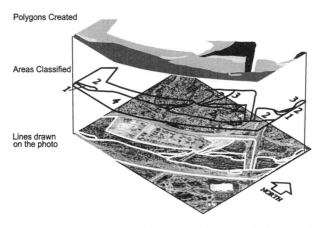

Figure 4 Vegetation mapping most often starts with an aerial photo. Polygons of vegetation types are drawn on the photo, and then field checked. The information is entered into a GIS system, where statistical analysis is run on the data and maps are produced.

4 illustrates a simplified vegetation mapping process. Polygons would be drawn around the more obvious vegetation types on the photo. A field check would be conducted and the polygons would be revised where necessary. The photos would be scanned and rectified (removing distortion from the photo), input into a GIS system and geographically referenced (placed in a map coordinate system, such as State Plane, UTM or latitude and longitude). The polygons would then be digitized on the photo image in the computer. Attributes such as vegetation type, height, density, and area would be added to the table associated with the vegetation polygons created

within the GIS system. A map of the area would be produced and various statistics run on the vegetation information.

Although the process is straightforward, there are some important points to consider. First, the vegetation polygons are drawn on the photo by the best judgement of the biologists. If an area is left in or out it is done so by someone's opinion, not because there is line drawn on the ground in the field. Strange as it may seem, this is a very hard concept for many, even those trained in science, to understand. If two people are given the same photograph the polygons would be drawn differently and the statistical outcome would not be the same. So is it even possible to measure vegetation changes? Yes, if the method used is consistent. That is, the methods used to establish the baseline data does not change in future surveys of that same area. In the next survey the polygons would not be redrawn, only the change in the vegetation type would be redrawn.

Second, when mapping things over time there is this problem of advancing technology. Suddenly when it is time to head out in the field and bring in the new information, all of the equipment used on previous surveys has become obsolete. The computer software has either been updated or completely replaced by a bigger and better product. This is a real problem and only foresight can help. Sometimes a computer may have to be set aside to store the old programs for future use. If that is not possible, then a decision will have to be made that may involve redrawing the entire project using the new software. The problem there is that the new statistics will vary from the old ones. If it is known that the study will continue over a two-year period or longer, much thought should be given on how to keep the data intact through all the changes in hardware and software.

Biological Evaluation

Determining if a habitat is suitable for special interest animal or plant species has begun to play a very important role in mapping natural resources. Building and construction of any type, be it roads, houses, hospitals, or schools can be slowed or stopped if it is deemed to encroach upon the habitat of a protected or special interest species. A GIS mapping project to evaluate a suitable habitat for these species can be set up similarly to the predictive model used in archaeology. That is, themes that represent habitat criteria can be overlain to determine the most suitable areas. First, a base map would be set up utilizing such themes as roads, section lines, contours, water courses, and the proposed project area. These themes can sometimes be obtained by contacting local governmental agencies, such as counties or cities.

In some instances a USGS map can be used, in which case a raster image, known as a DRG file, can be obtained either from the USGS or from private companies if the map is needed immediately. Next, themes that depict habitat type would be overlain on the base map. Most often fieldwork by scientists is required to gather information on these themes, and typically the fieldwork is done on an as-needed basis. For example, a detailed vegetation study might be required, call stations for a species of bird might be set up and mapped, or an area might be assessed for a particular plant. Since mapping this type of information is time intensive and costly to gather, only generalized maps exist for the public, and they should not be treated as conclusive if a construction project is dependent upon them. Once all the themes are gathered and entered into the GIS, a map can be produced showing the relationship of the project area to the habitat in question.

Mine Planning

GIS is playing a large role in the mining industry. Creating a mine plan of operation is a long and complex process employing GIS applications from both cultural and natural resources. Archaeological surveys, biological evaluations, vegetation studies, line-of-sight and intervisibility, jurisdictional waters studies, socioeconomic impacts to the community, are just a few of the areas that need to be investigated before mining can commence. By allowing for the rapid testing of many mine-plan models, a company can create a plan of operation that not only minimizes impacts but also maximizes efficiency. Issues such as traffic, noise, air quality, water quality, right-of-ways, grazing, and endangered species can all be mapped and analyzed. Furthermore, 3-D models can be created. Visual impacts and restoration plans can be designed and rendered in 3-D so that it can be viewed from any angle, giving the public a clearer understanding of how the mine may affect them or their community. These types of tools can help give a basis for discussion.

CONCLUSIONS

Two-dimensional mapping and data analysis are currently the main uses for GIS in the fields of cultural and natural resources, but interest is growing for the new 3-D modeling. Data sharing is becoming a viable means to acquire information within the field of archaeology as the information is being centralized into larger GIS programs in some states. Within natural resources data is more difficult to acquire, especially for individuals. GIS applications for cultural and natural resources will continue to expand as the technology advances. Imagination will be the primary limiting factor.

GIS has the ability to simplify the task, or cause so much complication as to boggle the mind. Typically, it takes more than one software package to complete a project and sometimes data end up juggling between three and four programs. Other more serious problems can arise from the fact that understanding the software and understanding mapping are two entirely different things. Issues of control and error can sabotage the best-laid plans. In order to begin any mapping task, one must have the software, hardware, data, and the understanding of how it all fits together.

ACKNOWLEDGMENT

Many thanks to Ken Kertell, Jerry Lyons, Kevin Wellman, and Dennis Dalton for their help and insights in writing this article. Special thanks to Jean L. Johnson for allowing me to use her 3-D model of the Hohokam pithouse.

REFERENCES

Carver, S. (1997). *"Innovation in GIS5: selected papers from the fifth national conference on GIS research UK (GISRUK)."* National Conference on GIS Research UK, Taylor & Francis, Bristol, PA, 239-243.

Ebert, J.I. and Hitchcock, R.K. (1980). *"Locational modeling in the analysis of the prehistoric roadway system at and around Chaco Canyon, New Mexico."* In Cultural Resource Remote Sensing, National Park Service, Washington, D.C., 167-207.

Worrall, L. (1990). *Geographic information systems: developments and applications.* Belhaven Press, New York, N.Y.

CHAPTER 11

Emergency Response (Disaster Management)

Emelinda M. Parentela and Shashi Sathisan Nambisan

Disasters typically strike with little or limited warning. However jurisdictions that have developed good emergency response plans are better able to respond to such disasters and emergencies than those that have not developed such plans. Preparing and implementing good emergency response plans require integration of a plethora of information. Information required for such purposes ranges from those pertaining to emergency response providers (location, capabilities, etc.) to the spatial and temporal distribution of various groups that could be potentially affected such as population sub-groups (elderly, children, hospital bound, etc.), and those related to the environment and the economy. Geographic information systems (GIS) based programs afford various capabilities to facilitate and automate analysis and planning for emergency response and its implementation. Several GIS based methodologies that can support such efforts are presented and illustrated in this chapter. A case study related to the potential transport of hazardous materials in the state of Nevada is used to illustrate these methodologies. Such analysis provides critical support for policy, planning, and operational decision making ranging from allocation of resources to response strategies.

INTRODUCTION

Natural and man-made disasters often result in fatalities and damages amounting to millions of dollars. A recent example is the El Nino 1998 that left devastating effects as a result of heavy rains, flooding, mudslides, and hurricanes. Natural disasters, like earthquakes, often come without warning. Others such as flooding, hurricane, tornado, mudslide, and volcanic eruptions can often be detected in advance, thus providing opportunities to prepare for such an emergency. Man-made disasters such as hazardous materials spillage and nuclear explosion are often accidental in nature, although sabotage is not uncommon. While these events may be uncontrollable, pre-event emergency response planning and preparedness, along with a well-planned and immediate response could minimize or mitigate, if not totally eliminate, their devastating outcomes.

The potential for a disaster, often with catastrophic outcomes, along with the need to ensure public safety and security, emphasizes the importance of evaluating both the risks posed by these events as well as the emergency preparedness of the affected community. Emergency preparedness is one aspect that has been recognized as crucial in disaster and risk management. The key aspects in disaster and emergency response planning involves the identification of critical areas, estimation of affected population, location of sensitive or difficult to evacuate population, evacuation shelter, response centers (e.g. police stations, fire stations, and hospitals), evacuation route, and warning/communication system. Analytical models and tools, such as those afforded by a geographic information system (GIS) program, are very useful in mapping out disaster areas, performing risk and emergency preparedness analysis, and facilitating disaster management. Because of the capabilities they afford to manage, analyze, and display spatial information, GIS-based programs and tools are particularly relevant for such purposes. The results of these analyses could be used in devising disaster preparedness and management/mitigation strategies. Furthermore, GIS programs also assist decision-makers in resource management and allocation for such programs.

LITERATURE REVIEW

The potential for a disaster may be measured in terms of *risk*, which is defined as the product of the probability of an event and its consequences, where an event can be described as any type of disaster. One of the most common applications of risk analysis is in making routing decisions such as in selecting evacuation routes, or routes for hazardous materials shipments. Risk minimization has been used as one of the objective functions in various route characterizations and analyses (Abkowitz 1993, List and Mirchandani 1991, Souleyrette and Sathisan 1994). The significance of local or route specific data for risk analysis is not to be overlooked in disaster management or emergency response activities (Madhavapeddi and Sathisan 1995).

Location-allocation models and decision-support systems have been developed for determining shelter locations (Sherali et al. 1991), for rural network evacuation (Radwan et al. 1985), and for logistics problems in the aftermath of an earthquake (Ardekani and Hobeika 1988). Rubec et al. (1998) and Washburn et al. (1998) present examples of GIS-applications for emergency response for disasters related to chemical or hazardous materials spills and emergency-911 systems, respectively.

The presence of an emergency response facility within an area could help reduce the consequence of an event in terms of attending to affected population and environment, timely evacuation, and containment of the area. The effectiveness of an emergency response unit depends heavily on response time or ability to respond immediately to an event, its capabilities, and resources. Emergency response preparedness is recognized as one way of minimizing the consequences of accidents after they occur and in minimizing community vulnerability. Abkowitz et al. (1991) and Parentela (1996) addressed various issues related to emergency response preparedness for disasters. They noted that emergency response is becoming a dominant area of concern and that risk aversion, risk equity, emergency response, and proximity of sensitive facilities are emerging as other factors to consider in risk analysis. Anders and Olsten (1990) and LaSarre et al. (1990) used GIS to assess risks and vulnerability of population by integrating emergency response times and capability. Abkowitz (1993) and Parentela et al. (1994) recognized emergency response and minimum emergency response time as a routing criterion.

APPLICABLE METHODOLOGIES

Various risk models include enumerative indices, regression models, network and distribution models, and probabilistic risk assessment models. Others include simulation models, subjective estimation, and over the last decade GIS-based methods and tools. Enumerative indices are based on a rating or scoring scheme. Regression models are the most common techniques and they employ historical data. Simulation modeling mimics actual events. Subjective estimation employs judgment of a panel of experts. GIS-based methods are relatively new techniques and they facilitate detailed evaluation and also help integrate spatial and temporal data. Such methods could be used to support all the aforementioned types of models - those ranging from subjective assessments to quantitative or probabilistic analyses.

Various capabilities or functions afforded by GIS programs may be used in performing the tasks required to support decision making. These include analytical tasks involving risk estimation, response time analysis, service area analysis, resource allocation, critical area identification, impact analysis, and emergency vehicle and evacuation routing. Some of the functions/capabilities afforded by most popular GIS programs include the following: address matching or geocoding, routing and allocation, Thiessen polygon (also known as Voronoi diagrams), buffering, dynamic segmentation, and overlay functions. These are discussed briefly next.

Address Matching/Geocoding

Geocoding generally refers to providing spatial coordinates (in two or three dimensions, and with a temporal component, if necessary) to objects. Address matching is the process of matching an object with its correct spatial address. Objects could be houses, schools, shelters, bus stops, etc. Various strategies are used to provide spatial coordinates. Geocoding is a four-step process. This includes preparing a reference theme or base map for geocoding, creating an address table that contains the addresses, matching the addresses, and creating geocoded points. Examples of such strategies include latitude/longitude, x/y, street address, route name/number along with milepost, or offset from known intersection. For emergency preparedness and disaster management key objects include locations of emergency responders such as police stations, ambulance, firefighters, hazardous materials response teams, and emergency response facilities (e.g. shelter, hospitals, and clinics). In case of evacuation, they include locations of sensitive and difficult to evacuate facilities such as adult care facilities, nursing homes, childcare centers, hospitals, schools, and prisons. Address matching refers to the processing of data with locational information about objects and matching them to a map (e.g. street network) with a known coordinate system.

Routing

Routing tools facilitate pathfinding within a network. Given an origin and a destination, various options to specify the objective function are typically available in

GIS programs (e.g. shortest path or quickest path). Additionally, a few programs or models developed from the basic GIS programs also permit introduction of simple or complex constraints such as points that must be visited during a trip. Such tools can be used to determine potential routes for evacuation purposes, or find the route that will minimize potential risk.

Allocation

Allocation is the process used to identify the maximum coverage that can be achieved from a location given specific constraints. Thus, in the allocation process, links of a network are assigned to a center until the maximum impedance limit is reached. It is very effective in evaluating response times of emergency response units - such as how much area or region can be covered by a unit stationed at particular location (base or center) within a specified time given specific travel times/ speed on links. Using the existing network, response time analysis could be performed to evaluate the response times given an existing condition (e.g. travel time, operating speed, time of day). This analysis uses travel time as an impedance measure. With the emergency response units serving as centers, the extent of coverage available for various desired response times can thus be evaluated.

Buffering

Buffering is a very common method of reclassification. Through the buffer command, a polygon is created around an object (which could be a point, line, or polygon). The region within the polygon represents the area that is within the tolerance specified for the buffer. For example, a buffer around a point will be a circle. Buffer analysis is extremely useful in performing proximity analyses. In emergency response or disaster management, it enables estimation of affected areas. Such areas may include exposed population and ecologically sensitive areas, where response facilities and/or shelters within the affected area are identified.

Thiessen Polygon (Voronoi Diagram)

A Thiessen polygon, named after the climatologist A.H. Thiessen, is also commonly referred to as Dirichlet diagram (region), Voronoi diagram, proximal zones or Wigner-Seitz cells (DeMers 1997, Tomlin 1990). It is used to determine the influence of point data or center, such as an emergency response station, to its proximate region. A Thiessen polygon for a center is the loci of points that are closer (in Euclidean space or as the crow flies) to the center than to any other center. Estimates of the service areas of emergency responders could be obtained using Thiessen polygon - which is the process of finding the proximal region of influence to be served by the responder (point). These polygons can be overlaid with other data themes for analysis - e.g. with a population data theme to estimate the population to be served by various response centers.

Dynamic Segmentation

One method of evaluating and combining multiple sets of linear attributes, such as a route network, is by dynamic segmentation. Events are related to the route system where the segments are not predefined. That is, the extent of segments could change based on the criteria used to define segments. Dynamic segmentation could facilitate risk estimation. For example, for estimating risks along potential hazardous materials shipment route, linear data such as accident rate, traffic volume, and hazardous material shipment quantity may be used as events that serve as a basis to define segments. Thus, it could be a very useful tool in identifying critical route segments as a part of emergency response planning.

GPS-GIS

Global positioning system (GPS) instruments may be placed on emergency response vehicles or responders for real time monitoring of their location. In simple terms, GPS receivers utilize real time data received from a constellation of satellites to identify the location of the receiver. This is a very simplistic description of what GPS receivers can do, and a detailed description of the capabilities and use of GPS units is beyond the scope of this chapter. Thus, data from GPS units can greatly assist the dispatchers in deploying and scheduling emergency response units and in disaster management.

EXAMPLE APPLICATIONS

Examples of the applications of GIS for emergency response and disaster management are provided in this section. It focuses specifically on assessing the emergency preparedness of a community when responding to a hazardous materials transportation incident.

Route Identification and Characterization

Yucca Mountain, Nevada, is currently being studied by the United States Department of Energy (DOE) as the only site for permanent disposal of high level radioactive wastes (HLRW) and spent nuclear fuels (SNF). If the decision is made to construct a permanent repository at Yucca Mountain, HLRW and SNF (currently stored at 76 nuclear power and some facilities operated by the U.S. federal government) will have to be shipped to the proposed disposal site. These origins are dispersed throughout the continental U.S. with a majority of them being located east of the Mississippi River. If the shipments were to be made by road (as opposed to rail), several potential routes exist within the state of Nevada.

For illustration purposes, consider that shipments arriving from the east enter Nevada on I-80 near Wendover. Figure 1 illustrates two potential highway routes that could be used to transfer the shipment from Wendover to Yucca Mountain. Further,

GIS-based tools have been developed to evaluate and quantify characteristics of alternative routes such as population or environmentally sensitive areas proximate to the route, geometric and operational conditions of the roadway, and so on. The emergency preparedness and potential risk along these transportation corridors are to be evaluated.

Location and Characteristics of Emergency Responders

Emergency preparedness is evaluated in terms of response times, and risk is measured in terms of affected population, accident potential, and the emergency preparedness of the area. For simplicity of analysis, the initial response to an incident is assumed to be provided by Police or Sheriff's stations (in reality, one would have to also consider fire stations and other emergency responders like ambulance providers).

The locations of police/sheriff stations with respect to the potential routes are shown in Figure 2. Also shown are Thiessen polygons for each station. The locations of police/sheriff stations can be represented as points. Each point is surrounded by a polygon (its Thiessen polygon), which represents the area or locality that is closer to the subject police/sheriff station than to any other station. The process of creating Thiessen polygons start by connecting each pair of neighboring points with an imaginary straight line. A line is used to bisect this imaginary straight line. The bisectors define the Thiessen polygon or the region that is most proximate to the subject point. The Thiessen polygon for a station simply represents the areas that are closer to the station (as the crow files) than to any other stations.

This could be combined with population data or network characteristics (e.g. cumulative centerline miles of roadway within a station's jurisdiction, traffic data, etc.) to perform comparative analyses of the resources deployed. For example, if the objective were to evaluate the emergency preparedness, then the area to be served by each of the emergency responders needs to be estimated. Using the concept of nearest neighbor policy - wherein the responder nearest to the location of an event will be expected to respond first regardless of jurisdiction - Thiessen polygons are used to determine the area coverage of individual police/sheriff stations (see Figure 2). The figure shows that some stations would have to serve a larger area. Other stations cover smaller areas, which are often found in cities or other densely populated areas. Clearly, the area distribution may have some serious implications in terms of response times.

Attributes could be coded for each station to characterize its capabilities and resources. For example, if any of the stations were specifically prepared to respond to hazardous materials incidents, this could be identified (in reality, it is typically fire stations that have such capabilities). Thus it would be possible to differentiate between the initial responders and those that have the special capability to respond to hazardous material transportation incident. This is shown in Figure 3 for Clark County, Nevada.

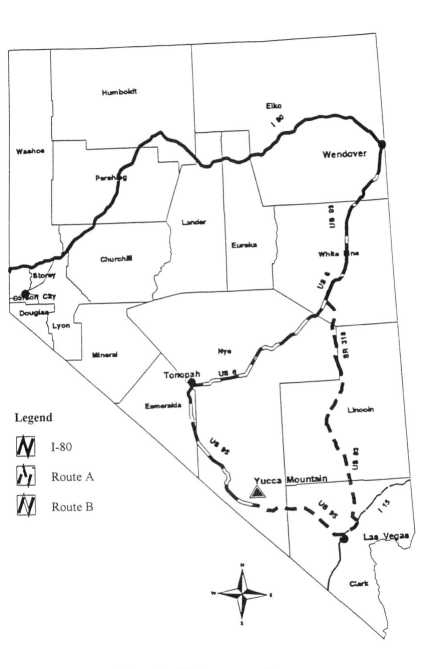

Figure 1   Highway routes of interest

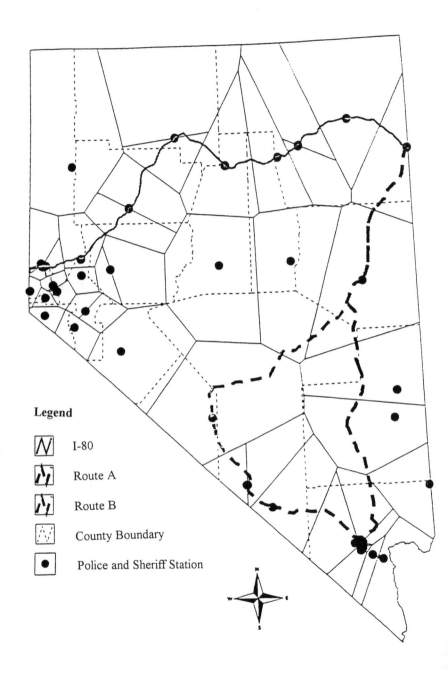

Figure 2 Thiessen polygons for police/sheriff stations

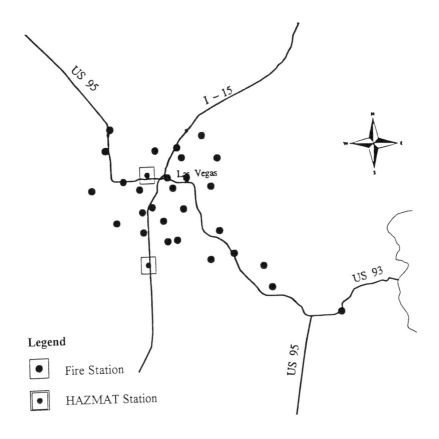

Figure 3   Locations of fire stations and hazmat stations in Clark County, NV

Response Time Analysis

Response times of 10 minutes, 30 minutes, and 60 minutes are evaluated using network analysis. The analysis to determine the response area is as follows. Each link in the network is assigned travel time impedance that is based on the speed limit. Travel speeds on the network are obtained from the link characteristics. For example, posted speed limits may be used for this purpose. Otherwise, if such attributes are not available, assumptions may be made - e.g. travel speeds of 55 mph for Interstate highways and U.S. routes and 25 mph for other types of road. Then, each police or sheriff station is used as an origin or center. Using features such as allocate, the extent of the networks around the stations that can be traversed within a specified response time can be determined. Beginning at the station, GIS process will add the travel times as each route spreads outward until the maximum capacity or impedance, represented as response time, along all the paths allocated to the station is reached. This is illustrated in Figure 4, which identifies the area that can be served by the station within a 10-minute response time. Similarly, analyses are performed for 30-minute and 60-minute response times and the results are presented in Figures 5 and 6, respectively.

The results of such analyses can be used to prepare for emergencies as well as to respond to disasters. For example, the results could be used to allocate resources to deploy a new unit at a new location or to deploy additional units at existing locations based on policies and management decisions. If, for example, the state's (or DOEs) policy was that the response time along potential transport corridors for shipments to Yucca Mountain were not to exceed 30 minutes, Figure 5 could be used to identify the sections of the network that did not conform to this policy. Then, decisions would have to be made to either deploy additional units at new locations or to not use the route for shipments. If a disaster did occur, the GIS-based system could be used to identify not only the closest responder, but also the nearest responder with capabilities to respond to the disaster (e.g. a major chemical spill). It can also identify other locations from where responders may be called without significantly compromising the capabilities at those locations.

In another analysis, recognizing the relatively smaller area covered by Clark County, response-time contours of 5 and 10 minutes were developed for the area. These are then overlaid on a coverage with population information and it is shown in Figure 7. The figure shows that route segments that traverse the more heavily populated areas can be reached by a responder within 10 minutes.

Another critical aspect of a disaster is evacuation - specifically the evacuation of sensitive and/or difficult to evacuate populations such as the elderly, children, sick, and prisoners. An inventory of the location of centers housing these population groups will greatly aid in the planning of evacuation strategies (such as determining evacuation routing and shelter locations). GIS-based tools were developed to identify locations of sensitive population within specified buffer zones (e.g. within a 10-mile corridor along a transport route), such as child-care centers. Similarly, tools were also developed to identify locations of emergency response facilities such as hospitals. A similar method may be used to identify the location of evacuation centers.

CONCLUSIONS

This chapter has illustrated a few GIS-based concepts and analytical techniques that are applicable to disaster mitigation and emergency response planning. Databases and results of GIS-based analyses could help identify critical segments or areas, evacuation routes, locations of emergency responders, degrees of preparedness of an area, and in improving current emergency response and disaster management practices.

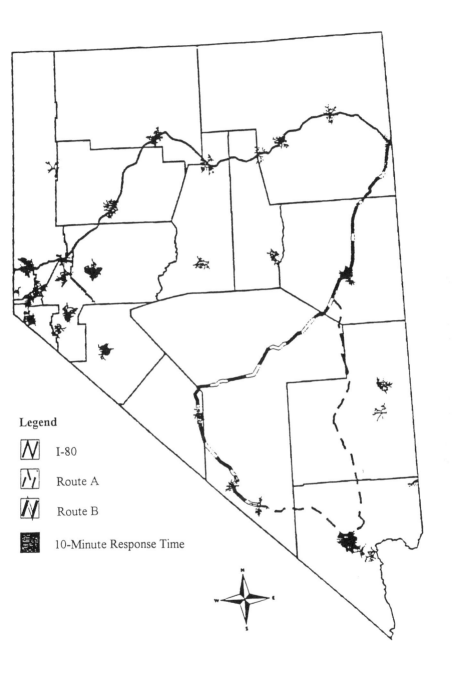

Figure 4 10-minute response times of police/sheriff stations

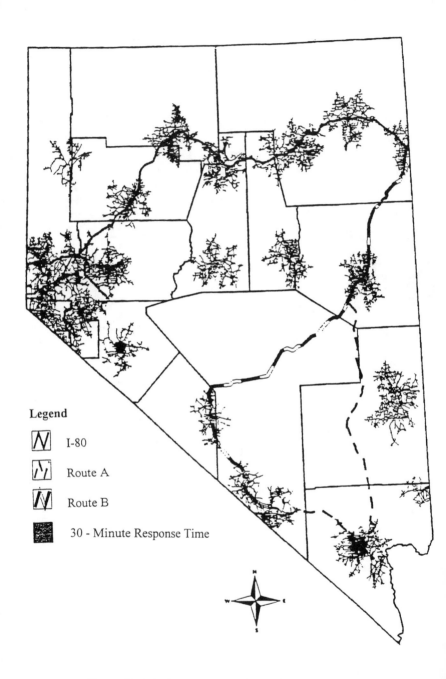

Figure 5 30-minute response times of police/sheriff stations

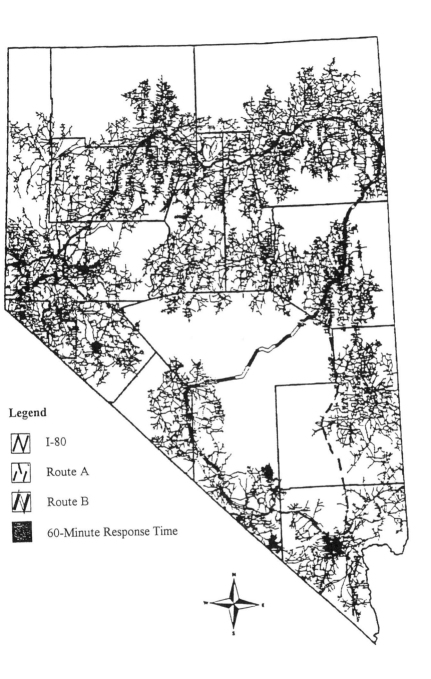

Legend

I-80	
Route A	
Route B	
60-Minute Response Time	

Figure 6   60-minute response times of police/sheriff stations

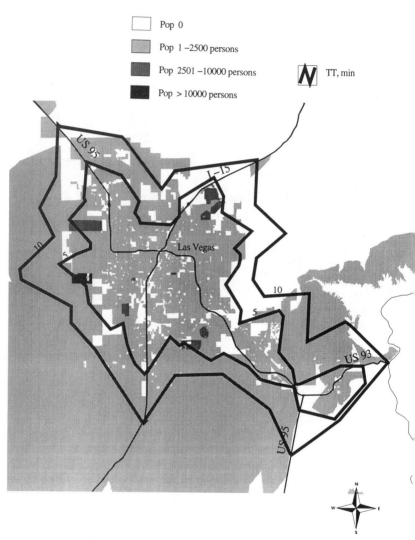

Figure 7   Response time and population distribution

ACKNOWLEDGMENTS

The authors gratefully acknowledge the support from various organizations and individuals. The State of Nevada Agency for Nuclear Projects and the United States Department of Energy provided some funding for work that served as a basis for this study. We thank the staff of various emergency responders located throughout

Nevada (in particular, Jim O'Brien of the Clark County Fire Department) for their support in the form of information and advice. We also thank several students and staff at the UNLV Transportation Research Center for their assistance.

REFERENCES

Abkowitz, M.D. (1993). "Working together to build a safer future." *Transportation of Hazardous Materials*, L. Moses and D. Lindstrom (eds.), Kluwer Academic Publishers, Boston, MA, 85-101.

Abkowitz, M.D., Alford, P., Boghani, A., Cashwell, J., Radwan, E., and Rothberg, P. (1991). "State and local issues in transportation of hazardous materials: toward a national strategy." *Transportation Research Record 1313*, TRB, National Research Council, Washington, D.C., 49-54.

Anders, C. and Olsten, J. (1993). "GIS risk analysis of hazardous materials transport." *In State and Local Issues in Transportation of Hazardous Materials: Towards a National Strategy. Proceedings of the National Conference of Hazardous Materials Transportation*, M. Abkowitz and K. Zografos (eds.), ASCE, Reston, VA, 248-261.

Ardekani, S.A. and Hobeika, A.G. (1988). "Logistics problems in the aftermath of the 1985 Mexico City earthquake." *Transportation Quarterly*, 42(1), 107-124.

DeMers, M.N. (1997). *Fundamentals of geographic information systems.* John Wiley & Sons, New York, N.Y.

Lasarre, S., Fedra, K., and Weigkrict, E. (1990). "Computer-assisted risk assessment of dangerous goods for Haute-Normandie." *In State and Local Issues in Transportation of Hazardous Materials: Towards a National Strategy. Proceedings of the National Conference of Hazardous Materials Transportation*, M. Abkowitz and K. Zografos (eds.), ASCE, Reston, VA, 281-295.

List, G.F. and Mirchandani, P. (1991). "An integrated network/planar multiobjective model for routing and siting for hazardous material and wastes. *Transportation Science*, 25(2), 146-156.

Madhavapeddi, K. and Sathisan, S.K. (1995). "On significance of route specific data for transportation risk analysis." *Proceedings of the 6th International High Level Radioactive Waste Management Conference and Exposition*, ASCE, Las Vegas, NV, 481-483.

Parentela, E.M. (1996). *A framework for modeling risk and emergency response in hazardous materials transportation.* Ph.D. Dissertation. University of Nevada, Las Vegas, NV.

Parentela, E.M, Burli, S., Sathisan, S.K., and Vodrazka, W.C. (1994). "Emergency response preparedness analysis for radioactive materials routing." *The 5th Annual International High Level Waste Management Conference and Exposition*, Las Vegas, NV, 22-26.

Radwan, A.E., Hobeika, A.G., and Sivasailam, D. (1985). "A computer simulation model for rural network evacuation under natural disasters." *ITE J.*, 55(9), 25-30.

Rubec, P., Norris, H., and LaVoi, T. (1998). "New technologies for emergency response: testing a prototype system in Florida." *Geo Info Systems*, November, 20-26.

Sherali, H.D., Carter, T.B., and Hobeika, A.G. (1991). "A location-allocation model and algorithm for evacuation planning under hurricane/flood conditions." *Transportation Research*, 25B(6), 439-452.

Tomlin, C.D. (1990). *Geographic information systems and cartographic modeling.* Prentice Hall, Englewood Cliffs, N.J.

Souleyrette, R.R. and Sathisan, S.K. (1994). *GIS for radioactive materials transportation.* Special Issue of Microcomputers in Civ. Engrg. J., 9, 295-303.

Washburn, D., Beaumont, J., and Gruber, S. (1998). "Orange County, New York's enhanced-911 system helps residents breathe easier." *Geo Info Systems*, November, 28-34.

CHAPTER 12

Environmental Assessment of Transportation-Related Air Quality

Wayne Sarasua, Shauna Hallmark, and William Bachman

The capabilities of geographic information systems (GIS) are currently being exploited in the environmental assessment of air quality. In particular, GIS tools have been used to assist planners, engineers, and researchers in the spatial and temporal quantification and analysis of automobile exhaust emissions, a major contributor to air pollution. This chapter focuses on the varied uses of GIS in the environmental assessment of air quality impacted by transportation sources. The chapter uses a four-tiered approach to explain the different levels of integration of GIS in environmental air quality assessment ranging from spatial and attribute database management tools, such as those used for input and thematic display, to sophisticated modeling uses of GIS. The chapter illustrates several real world examples to reinforce the concepts presented.

INTRODUCTION

Air quality and other environmental issues have influenced transportation engineering, planning, and research for a number of decades. The relationship between urban air quality and transportation is complex. The transportation sector contributes between one-third to one-half of ozone precursors, one-half of nitrogen oxides (NOx), and between two-thirds and nine-tenths of carbon monoxide (CO) in urban areas (Anderson 1995, USDOT 1993). Areas in non-attainment for transportation-related emissions, such as CO, are required to show timely reductions in emissions and demonstrate that new transportation projects will not lead to new violations (Washington 1995). Accordingly, transportation planning, engineering, research and new investment funding are often tied into air quality analysis, forcing researchers and agencies alike to find improved methods to carry out transportation studies.

Air quality in and of itself is spatial. The formation, release, and effects of transportation-related emissions are spatially and temporally dependent. A geographic information system (GIS) is especially well suited to air quality analysis and provides unique opportunities for managing, summarizing, analyzing, and modeling this complex connectivity between transportation and urban air quality. As a result, given the federal regulatory reporting requirements for meeting air quality standards and its current importance to state and local officials, many are using the powerful tools found in GIS to manage and report the vast amount of information needed by consultants and public agencies.

The National Cooperative Highway Research Program (NCHRP) *Report 359* studied GIS in an effort to define its potential uses by transportation agencies (NCHRP 1993). The document, which presents a comprehensive overview of GIS technology, its potential role for serving the needs of a variety of agencies, and strategies for successful implementation, states that the impact of GIS-T is profound and if exploited fully would become an integral part of information processing environments. Further, it states that solution of environmental and economic development issues requires the interaction and sharing of geographically referenced data from all levels of government as well as other sources of data. Consequently, GIS is the mechanism that ties transportation and air quality, as well as the tie that brings jurisdictions together.

GIS has many applications for assessing ambient air quality or emissions, which can be tailored to an agency's needs and resources. Consequently, a multi-level approach for using GIS as an air quality analysis tool exists. Some agencies use a GIS only for managing the input to air quality models and the development of inventories. Others have developed sophisticated models capitalizing on GIS functions from low-end uses such as data storage all the way up to sophisticated spatial analysis. Figure 1 shows a pyramid of the various levels of integration of GIS use in the environmental assessment of air quality. As you move up the pyramid, the level of sophistication increases. Many of the different types of GIS/air quality links are used either alone or in different combinations, which may be tailored to an agency's needs and resources. The different levels of integration shown in Figure 1 are: spatial and attribute data management, spatial summarization, spatial analysis, and modeling 'what-if' scenarios.

SPATIAL DATA MANAGEMENT

The spatial data management level of integration includes using the GIS to input, store, retrieve, and display air quality related data, such as the location of pollution sources and monitoring stations. Many GIS/air quality applications stop at this lowest level. Even without high-end analysis and modeling applications, a GIS is well suited

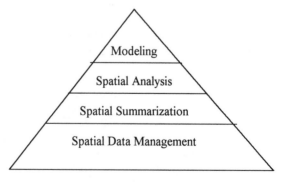

Figure 1 GIS integration levels

as a storage and maintenance platform for spatial data and associated attributes. The following sections describe how the spatial data management capabilities of GIS have proven beneficial for environmental air quality assessment.

Data Input

Data input involves encoding data and writing them to a database. Both the geographic data necessary to define where planimetric (e.g. roads), cadastral (e.g. city or land ownership boundaries), or other features occur, and attributes associated with those features need to be considered when using a GIS for air quality assessment. Data input occurs in a variety of ways, including geocoding and manual input.

Batch Geocoding

Batch geocoding can be done by processing data and assigning coordinates based on specific location identifiers contained in the data such as addresses, zip codes, or traffic analysis zones (TAZ). Batch geocoding of data for use in air quality assessment has been done extensively. Using address matching, millions of vehicle registration records can be geocoded along with critical attributes such as vehicle make, model, or engine type. This data are beneficial to activity-based emissions modeling because emission-producing activities vary significantly depending on vehicle characteristics. GIS analysts have also used address geocoding in the assessment of automobile inspection and maintenance records. Spatial patterns regarding malfunctioning vehicles have been spatially correlated with income levels. Figure 2 displays geocoded vehicle registration records.

Traffic analysis zone data are fundamental to four-step travel demand forecasting models. The U.S. Environmental Protection Agency's mobile emission assessment system for urban and regional evaluation (MEASURE) is a research emissions model that runs entirely in a GIS (Bachman 1997) and is discussed in detail in the modeling section of this chapter. MEASURE uses trip generation data that are batch geocoded to TAZ to estimate the number and location of engine starts.

Another application of batch geocoding is travel survey data. Georgia Tech looked at travel survey data for Atlanta and Seattle that were batch geocoded to TAZ. An estimated average soak time (engine off time) by trip type and location was developed that helped researchers explore the differences in the spatial patterns of cold start emissions.

Manual Digitizing

Manual digitizing can be useful in almost any GIS application including air quality assessment. It is common to digitize planimetric features from high quality aerial photos. In Atlanta, air quality issues are currently playing an important role in the success or failure of the Atlantic Steel Development project just north of downtown Atlanta. For this project, the roads and land uses (e.g. day care centers, parks, schools) were manually digitized from low altitude aerials to ensure spatially accurate

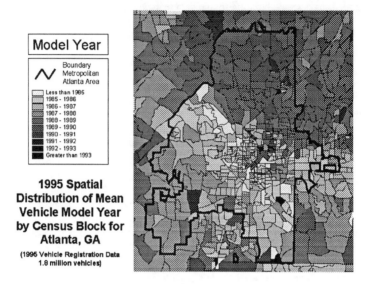

Figure 2 Geocoded vehicle registration information

inputs for air quality models. Computer aided design drawings of the development were combined with digitized surrounding facilities and roads (Figure 3), all of which were used in the estimate of CO concentrations.

Remote Sensing

Satellite and aerial remote sensing are a means by which large amounts of planimetric data can rapidly be input to a GIS. Much of the air quality assessment work that has been done relies on vector data. Using 'heads-up' digitizing techniques, raster image data can be traced into vectors right on the GIS display.

GPS

GPS has proven to be beneficial for geocoding point, line, and polygon data for a variety of GIS applications including those involving air quality assessment. Precise geocoding of receptor locations can be accomplished quickly and efficiently using GPS. One unique application of GPS is the collection of roadway grades using a specialized attitude GPS that can kinematically measure heading, grade, and roadway banking simultaneously (Awuah-Baffour et al. 1997). High engine loads resulting from steep roadway grades negatively impact automobile emissions. Recent studies indicate that under high-load events, instantaneous vehicle emissions increase by up to one order of magnitude for hydrocarbons (HC) and NOx, and up to two

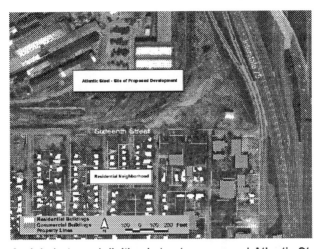

Aerial photo and digitized structures around Atlantic Steel
in Atlanta, Georgia.

Figure 3 'Head-up' digitizing of aerial photos

orders of magnitude for carbon monoxide. In particular, geometric conditions, such as grades greater than 3%, promote high load events for vehicles at high speeds (Cicero-Fernandez 1997). Because of its effect on emissions, grade data will likely be an integral part of future mobile emission models.

Figure 4 illustrates a temporal GIS system used for analyzing exhaust emissions of specially equipped vehicles (Sarasua et al. 1996). In this system, GPS is the primary data collection tool of vehicle position. The figure shows several things. First, it shows the location of a vehicle at a given instance. Second, it is able to thematically display the current value of an engine parameter or emission value that time (in this case throttle position). Third, it provides a dyanamic bar graph of several engine parameters/emission values for comparison purposes. Additionally, the dynamic system can be used to link dynamic information with underlying static information such as grade or elevation.

Data Retrieval

Forward Data Display. Forward data display is a method of querying data by selecting a spatial entity and displaying that entity's attributes. In MEASURE, Georgia Tech's GIS-based mobile emissions model, which is discussed in a later section, the user can select a road or zone and retrieve the total, mode-specific, pollutant-specific, and/or hour-specific emission or vehicle activity estimate.

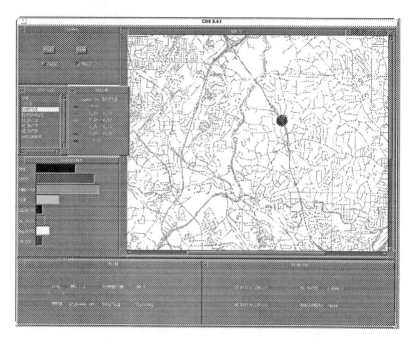

Figure 4 Temporal GIS using GPS data input

Backward Data Capture. Backward data capture is the opposite of forward data display. The user enters an attribute, and the spatial entities associated with it are displayed. For example, the user can type intersection cross streets for use in microscale air quality analysis and the intersection will be displayed on the map.

Conditional Queries. Hallmark and O'Neill (1996) integrated GIS with an intersection air quality analysis program for microscale air quality analysis within GIS. Figure 5 shows a conditional query displaying source and receptor locations with CO levels violating the one-hour national ambient air quality standards of 35 ppm. Once emission data are stored in a GIS, conditional queries make the GIS a powerful assessment tool. For example, traffic engineers with a budget for reducing adverse emissions in their district, could search the GIS for intersections with the most serious congestion and then develop site-specific strategies, such as improved signal timing, for emission reduction.

Data Display

The main display capability of a GIS is creation of thematic maps, which are capable of displaying attribute information using colors, scaled symbols, graphic charts, and associated legends. Figure 6 shows a thematic map illustrating the cold start emissions by Census Block for the Atlanta, GA metropolitan area.

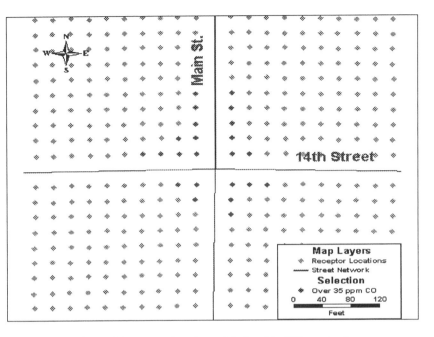

Figure 5 Receptor locations violating 1 hr CO standards

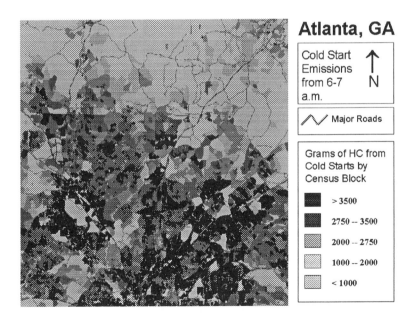

Figure 6 Thematic output of grams of HC by census block

Data Storage

Efficient storage of spatial data and attributes is essential for swift retrieval. Air quality related data, such as source locations, can be efficiently stored in conventional GIS using an arc-node data model. A direct linkage can be developed between geographic points/nodes, arcs, or polygons and the associated attributes. If data can be directly related to a specific arc or segment, such as average daily traffic volumes, the classic GIS arc-node data model can be used. However, a problem presents itself when trying to store continuous data. For example, grade data can vary continuously over a link between two nodes. To store this data accurately, the link needs to be segmented into smaller links. Overlaying other linear data such as speed and lane geometry can require breaking the segments further. Dynamic segmentation used in conjunction with a milepoint linear referencing scheme provides a more efficient means of storing varying link data. See Chapter 5 for more details.

Dynamic segmentation is a mechanism for segmenting a road network into sections without physically breaking links and adding nodes. Instead, linear data can be stored based on a *from* and *to* mile point. Using dynamic segmentation, hot stabilized vehicle emissions can be modeled microscopically along an arc. Dynamic segmentation can also be used to allocate start emissions to the road network as opposed to using a less precise 'puff' method that allocates start emissions uniformly to a zone. With dynamic segmentation, emissions resulting from engine starts can be allocated to the network based on travel time. Further, dynamic segmentation can be used to segment arterial links into areas of homogeneous modal vehicle activity. Mounting evidence indicates that vehicle emissions are highly correlated with vehicle activity and that instances of high loading such as 'hard' accelerations cause disproportionate rates of emissions as compared to normal stoichometric vehicle operation (Guensler 1993, LeBlanc et al. 1994, Kelly and Groblicki 1993). Dynamic segmentation offers a built-in method to divide roadway links into modal activity zones including acceleration, deceleration, idle, and cruise zones based on distances from signalized intersections and other road characteristics so that activity specific emission rates can be applied.

Data Transform

Transforming data from different formats or projection systems is routine in a GIS. Forecasted travel demand volumes can easily be imported into a GIS from four-step modeling networks if there is a one-to-one correspondence between links. A correspondence table can be used to transfer data between differing networks. A challenge in air quality assessment is dealing with the different scales that are used for different applications. Most ozone photochemical modeling occurs for very large areas (e.g. an entire state or group of states), while emission inventory modeling occurs around non-attainment MPO regions. Microscale modeling is site specific. Moving between various levels of detail requires some knowledge of transformations and the potential misrepresentations that can occur.

SPATIAL DATA SUMMARIZATION

Some GIS fundamentalists distinguish between spatial summarization of data and spatial analysis of such data. In this discussion, spatial data summarization refers to basic functions for the selective retrieval of spatial information within defined areas of interest, and the computation, tabulation, or mapping of various basic summary statistics. The types of operations within this category include spatial operations (spatial queries and spatial aggregation), geographical overlay (line intersection, point-in-polygon, line-in-polygon, and polygon overlay), and the creation of buffers.

Spatial Operations

Spatial Queries. A spatial query is a special type of conditional query where the results of a query must meet a specified boundary condition. Identifying residents within a mile radius of a point source would be an example of a spatial query.

Spatial Aggregation. Spatial aggregation involves combining spatial entities into larger ones based on common attributes. One important factor influencing emission estimates for an area is vehicle subfleet characteristics. For instance, higher income neighborhoods typically have newer cars that operate cleaner than older vehicles of the same size. Using vehicle subfleet data matched to TAZ or Census blocks, GIS can spatially aggregate areas with similar subfleet characteristics to create geographic boundaries for homogenous fleet groups. Likewise, census blocks can be aggregated to larger polygons in GIS using spatial aggregation and the attributes directly transferred, added, or proportioned to the larger polygons based on their original areas.

Geographical Overlay

Line Intersection (Line-on-Line). It may be desirable to combine line data that exist on different layers onto a single layer. One example would be to overlay roadway grade data stored by milepost with speed and average daily traffic data stored by link into homogeneous segments. Emission factors could be applied on a link basis to determine link by link emissions estimates.

Point-in-Polygon. Point data are a significant portion of the data used for air quality assessment and is often allocated to larger polygons to facilitate emissions analysis. This point-in-polygon procedure can be used to determine zonal fleet distributions by overlaying zones (e.g. TAZ) and address. The analysis results in the ability to create zonal distributions of important emission-specific vehicle technologies.

Line-in-Polygon. One of the goals in air quality modeling and conformity analysis is to aggregate data used to calculate emissions to 4-km gridcells for use in a photochemical model, such as the urban airshed model (UAM), to determine an overall emissions estimate for an area. One of the critical data elements in calculating hot stabilized emissions is gridded vehicle miles traveled (VMT) estimates. VMT data are stored as line data in most GIS. Using a line-in-polygon procedure, the linear VMT data can be aggregated to grid cell polygons. This has become a very common (and useful) application of GIS. Once the VMT data are in a grid cell format, emission factors developed using the US EPA's MOBILE software can be applied to the VMT grid cells to provide a gridded estimate of emissions ready for subsequent modeling.

Polygon-on-Polygon. MEASURE predicts cold starts at the TAZ level using the zonal origin data. The UAM requires that hourly emissions be gridded to a user-defined grid cell size. A polygon-on-polygon operation provides an automated means to grid the cold start estimates. Once gridded, the cold start, hot stabilized (based on VMT), and enrichment events can be added and an overall grid cell estimate made.

Buffering

Buffering can be used to determine the number of trips originating within a specific distance of a major street segment to find the percent of vehicles along the link that will be in cold-start mode. It can also be used to determine the number of non-roadway sources of transportation-related emissions located near a particular transportation facility. For example, in microscale modeling the number of fast-food restaurant and bank drive-thrus or parking near a major intersection can be determined and included as additional sources of transportation pollutants.

SPATIAL ANALYSIS

The next level of integration between GIS and air quality assessment is using the GIS for spatial analysis, which investigates patterns in and between spatial data. In most cases, these techniques are limited to the research side of air quality assessment and are not required for currently established approaches to conformity and inventory modeling. Spatial analysis methods can be purely deterministic or can address the inherent stochastic nature of patterns and relationships. Classical deterministic analysis techniques are among the most important uses of GIS for analysis in a transportation application. Network analysis, routing, location/allocation modeling, 3-D modeling and projection, and cartographic algebra are examples of deterministic analysis methods. Statistical spatial analysis includes spatial autocorrelation, which is concerned with exploring spatial covariance structure in attribute data, i.e. whether and in what way adjacent or neighboring values tend to move together. To a lessor extent, geostatistical and spatial econometric modeling has been linked to GIS activities but little has been done with regard to air quality assessment.

Network Analysis and Routing

Air quality analysis relies heavily on travel demand forecasting networks for forecasted annual daily traffic and VMT estimates. Travel demand forecasting relies heavily on routing algorithms to provide travel times necessary for distributing trips and travel paths used in the assignment of these trips. While most travel demand forecasting is done outside of GIS, some GIS software have fully integrated network analysis and travel forecasting capabilities.

Another application of routing in air quality assessment is allocating intrazonal trip emissions to roadway links. One method used for analyzing intrazonal activity is to use trip origins to estimate the number of vehicles operating on the road, and then take the average network distance (from the census block centroids to the closest modeled road segment) and calculate average intrazonal travel time.

Optimal Location

Current GIS with extensive network analysis capabilities can optimally locate facilities to most efficiently serve an area, such as locating a school to minimizing the walking distance of students. There is significant potential for using the optimal location capabilities of GIS in air quality assessment. Once better inventory and photochemical models are available, it will be possible to locate new facilities based on their potential contribution to ambient emissions. For example, placing a new manufacturing plant in one location may impact regional air quality less than placing it in another. Given average wind patterns and estimated concentrations of other pollutants, GIS could be used to optimize a facility's location based on its expected pollutant contribution.

Three-Dimensional Analysis

As knowledge regarding ozone formation and atmospheric mixing improves, and as better regional inventory data are available, it is likely that 3-D spatial modeling will predict pollutant concentrations at various altitudes. There has been significant work completed in modeling pollutant dispersion from stationary sources (i.e. stacks). While this research was not initially conducted in a GIS environment, GIS is rapidly becoming the preferred medium due to the ease of making connections between 3-D spatial layers. This type of modeling is extremely data intensive and the most widely used software packages are not designed to handle this type of analysis. Similar to 3-D modeling is surface analysis. Surface analysis uses a 3-D shell to evaluate slopes and aspects. Air quality analysts have used this technique to identify the relative change of emission production across space and time. Three-dimensional modeling of emissions in Atlanta, GA is shown in Figure 7.

MODELING

The highest level of GIS use is where the data management, summation, and analysis capabilities are combined into a modeling application that can be used to answer 'what if?' questions such as identifying the traffic impacts of different land use scenarios or predicting air quality during a major event. The next sections provide

GIS-based air quality modeling, including analytical modeling, microscale air quality assessment, and regional emission inventory modeling.

Analytical Modeling

Point Source Air Quality Modeling

Osborne and Stoogenke (1991) found pcAir-1 (GIS-based air toxics management system) useful in managing the large amounts of data analyzed and reported for point source air-quality models. Information on emission sources, receptor locations, population, and topography was stored in the GIS and extracted to run air-quality models. Model output also could be input to the GIS. A GIS was useful in this study as it offered features such as mapped references, layered data storage, custom map creation, and integration of data from disparate sources including output from dispersion models and manipulation of large datasets. Database information used in the study included geographic location of pollution sources and attributes such as operating hours or stack characteristics, output from dispersion models, and chemical-by-chemical reference criteria for comparison to federal and state standards.

The integrated GIS and air-quality model allowed passing of geographic source data for model input requirements and integrating air-quality model output within a GIS format. Model output of pollution concentration overlaid with existing

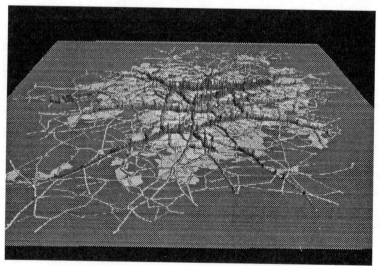

Mobile Source NOx Emissions (z-value) 7-8 AM in
the Atlanta Non-attainment Region

Figure 7 Three-dimensional modeling of CO emissions

geographic data allowed evaluation of both existing air pollution impacts and 'what if' scenarios. Other analyses included functions such as concentration information overlaid with other data to determine population exposure or cancer rate incidences.

Disaggregate Modeling

Othofer et al (1995) developed an approach to predicting location specific emission production estimates for changing control strategies. Instead of developing estimates using detailed location-specific emission producing activities and emission rates, they disaggregated large zonal estimates using emission-producing activities. The advantage of this approach is its simplicity and its straightforward recognition that the data needed to predict emissions at smaller levels does not exist or the relationships are undefined.

Microscale Air Quality Assessment

Regional air quality analysis includes estimates for principal transportation-related air pollutants including CO, HC, NOx, lead, and photochemical oxidants (ozone). However, carbon monoxide is usually the only pollutant that requires detailed microscale impact analysis because of the immediate health risks associated with exposure to high concentrations of CO. Transportation-related analysis is particularly important since mobile source emissions are the most significant source of CO. A GIS provides a number of benefits to microscale air quality analysis. A GIS is useful for spatial allocation of mobile sources, determining spatial relationships between emissions production, wind patterns, and evaluating pollution concentrations at specific locations.

GIS for Carbon Monoxide and PM_{10} Compliance

Souleyrette et al. (1991) undertook a microscale transportation air-quality study to determine compliance for carbon monoxide and PM_{10}, the bulk of which are from vehicle emissions, in the Las Vegas metropolitan area. The Las Vegas study focused on the relationship between vehicle miles traveled, refueling station locations, wind patterns, and CO concentration levels. Standard regression analysis was used to estimate the spatial correlation between refueling stations or VMT and air-quality levels.

To estimate the relationship between VMT and air-quality, VMT locations were related to corresponding CO contour values from the GIS and regression was used to determine spatial correlation among variables and degraded air-quality. The researchers concluded that integration of air-quality analysis and GIS was important because of the spatial nature of air-quality analysis and that traditional transportation air-quality analysis may not account for the spatial and temporal aspects of transportation and environmental conditions contributing to air-quality.

Linking GIS with Microscale Air Quality Model

Hallmark and O'Neill (1996) describe a prototype microscale air quality model that links CAL3QHC, a mobile source intersection air quality analysis model, with a GIS for local transportation-related air quality analysis. CAL3QHC uses average speeds, traffic volumes, signal characteristics, and synthetic queuing algorithms with average speed and idling emission factors from MOBILE to determine emission release from signalized roadways. The main output of CAL3QHC is pollution concentrations at specific spatial locations based on actual emissions released as well as wind speed and direction. Coupling the model with a GIS allows spatial analysis between mobile sources and locations of interest.

Spatial information such as street coordinates with accompanying attributes, such as traffic volumes or signal timing, and coordinate location of user specified receptor location are output from the GIS to a format compatible as input to CAL3QHC. These data are used to run the air quality model, which produces output in the form of pollution concentrations at receptor locations. Locationally referenced concentrations can be input into the GIS for hot-spot identification, estimation of contributions of off-road mobile sources, and impact analysis. Use of the spatially referenced concentrations can be used for generation of contours, classification, thematic analysis, and point-in-polygon and polygon-on-polygon overlay for various analytic purposes. In particular, the GIS-based model allows evaluation of specific pollution concentrations in the immediate vicinity of street segments as well allowing evaluating what-if scenarios for proposed land use or operational changes that lead to impacts on surrounding street segments. The GIS base for the model also provides valuable abilities such as data storage and manipulation.

Regional Emission Inventory Modeling

CAL-MoVEM

Bruckman et al. (1991) describe the use of GIS in developing gridded, hourly estimates of emissions. They also developed a model called CAL-MoVEM that used GIS to assist in developing mobile source estimates for input into photochemical models. The main function of the GIS in their model was the spatial aggregation of travel demand forecasting model features into a grid. They used spatially defined vehicle mixes by trip purpose, temporal factors, hourly temperatures, trip volumes, trip speeds, and modal percentages as inputs. The spatially defined inputs were combined with EMFAC7E emission rates to produce gridded hourly estimates of pollutants.

The work was accomplished as part of an ozone level study in the San Joaquin Valley in California. Zonal estimates were allocated to TAZ centroids that were re-allocated to grid cells. Link estimates were allocated to nodes and re-allocated to cells. The use of points to represent these features did not take full advantage of the spatial structure provided by the original input data. TAZ that spanned more than one grid cell were only allocated to a single grid cell dependent on where the centroid fell.

This strategy would require the use of grid cell sizes comparable to large TAZ, which are 30-40 square km for some metropolitan areas.

MEASURE

Georgia Tech researchers took Bruckman's work a few steps further and developed the GIS-based modal emissions model, MEASURE. The major differences between the two are: (1) the model includes modal emission rates as well as MOBILE emission rates, (2) user-defined grid cells, (3) an improved spatial aggregation technique, and (4) the inclusion of local road emissions. Modal emission rates are designed to estimate emissions for specific vehicle activities (idle, cruise, acceleration, and deceleration) and vehicle technology combinations (cold and warm engine starts, hot-stabilized, and enrichment). These modes of vehicle activities have different spatial and temporal characteristics. Over a region, engine start emissions are estimated for census blocks, and hot-stabilized and enrichment emissions are estimated for road segments (intersection to intersection). Emissions from the zones and lines are directly aggregated into user-defined grid cells by completing polygon-on-polygon and line-on-polygon spatial summarization. The main advantage of MEASURE is that it allows users to model a wide range of strategies that may have an effect on emissions (i.e. signal timing and high-occupancy vehicle lanes).

CONCLUSIONS

Air quality analysis is inherently spatial. Where and when pollutants are produced and released are as important as the total emission inventory for an urban area. Spatial proximity to populations is especially important in microscale air quality analysis, such as the evaluation of concentrations of CO and the resulting exposure rates. Consequently, the spatial characteristics of air quality give rise to a natural link between GIS technology and air quality analysis. In particular, GIS has numerous defining features, which lend itself to efficient air quality analysis. GIS allows for efficient storage and retrieval of spatial data such as street network information and the positions of point sources and related attributes. The manipulation of these data through spatial aggregation and spatial overlay within GIS provide an enhanced capability for performing air quality modeling. The thematic mapping capabilities of GIS makes it well suited for creating well designed descriptive maps that can be easily understood by decision makers. This chapter has described current and potential applications of GIS to the environmental assessment of air quality. While GIS has proven useful in the environmental assessment of air quality, its future role will likely increase with the advent of fully integrated GIS-based emission modal air quality models.

REFERENCES

Anderson, J.F. (1995). "Cleaner alternative fuels for fleets: an overview." *Transportation Research Record 1472*, Transportation Research Board, Washington, D.C., 17-25.

Awuah-Baffour, R., Sarasua, W.A., Dixon, K.K., Bachman, W., and Guensler, R. (1997). "GPS with an attitude: a method for collecting roadway grade and superelevation data." *Transportation Research Record 1592*, Transportation Research Board, Washington, D.C., 144-150.

Bachman, W. (1997). *Towards a GIS-based modal model of automobile exhaust emissions*. Doctoral Thesis, School of Civil and Environmental Engineering, Georgia Institute of Technology, Atlanta, GA.

Bruckman, L., Dickson, R.J., and Wilkonson, J.G. (1992). "The use of GIS software in the development of emissions inventories and emissions modeling." Air and Waste Management Association, Pittsburg, PA.

Cicero-Fernández, P., Long, J.R., and Winer, A.M. (1997). "Effects of grades and other loads on on-road emissions of hydrocarbons and carbon monoxide." *Journal of the Air and Waste Management Association*, 47, August, 898-904.

Guensler, R. (1993). "Data needs for evolving motor vehicle emission modeling approaches." *Transportation Planning and Air Quality II conference*, ASCE, New York, N.Y.

Hallmark, S. and O'Neill, W. (1996). "Integrating geographic information systems for transportation and air quality models for microscale analysis." *Transportation Research Record 1551*, Transportation Research Board, Washington, D.C., 133-140.

Kelly, N.A. and Groblicki, P.J. (1993). "Real-world emissions from a modern production vehicle driven in Los Angeles." *Journal of the Air and Waste Management Association*, 43, October, 1351-1357.

LeBlanc, D.C., Meyer, M.D., Saunders, F. M., and Mulholland, J.A. (1994). "Carbon monoxide emissions from road driving: evidence of emissions due to power enrichment." *Transportation Research Record 1444*, Transportation Research Board, Washington D.C., 126-134.

National Cooperative Highway Research Program. (1993). *Adaptation of geographic information systems for transportation. NCHRP Report 359*, Transportation Research Board, National Academy Press, Washington D.C.

Orthofer, R. and Loibl, W. (1995). GIS-aided spatial disaggregation of emission inventories." *The Emission Inventory: Programs and Progress*; Air and Waste Management Specialty Conference. Research Triangle Park, NC.

Osborne, S. and Stoogenke, M. (1991). "Improved management and analysis of industrial air emissions using a geographic information system." *Proceedings of the Annual Conference of the Urban and Regional Information Systems Association*. pp. 54-65.

Sarasua, W.A., Jia, X., Bachman, W., Awuah-Baffour, R. and Kimbrough, S. (1996). "Using a dynamic GIS to visualize and analyze mobile source emissions." *Proceedings of the GIS-T Symposium*, Kansas City, MO.

Souleyrette, R.R., Sathisan, S.K., James, D.E. , and Lim, S. (1992). "GIS for transportation and air quality analysis." *Proceedings of the National Conference on Transportation Planning and Air Quality*, ASCE, 182-194.

USDOT. (1993). *Clean air through transportation.* United States Department of Transportation and Environmental Protection Agency, Washington, D.C.

Washington, S. and Guensler, R. (1995). "Statistical assessment of vehicular carbon monoxide emission prediction algorithms." *Transportation Research Record 1472*, Transportation Research Board, Washington D.C., 61-68.

CHAPTER 13

Program Evaluation and Policy Analysis

Lisa DeLorenzo

Geographic information systems (GIS) are quickly becoming an important tool for policy analysis and program evaluation in urban administration and social science. This chapter explores the use of GIS in the analysis of neighborhood stability and evaluation of Housing Conservation District program in St. Louis City. There are three key benefits of using GIS in this type of analysis. First, thematic mapping permits the visualization of data clusters that may be associated with certain public policies. In addition, statistical procedures can find the absence or presence of clustering. The visual presentation of data in map form is particularly powerful for both influencing public policy and conveying information in a non-technical way to a variety of policy stakeholders. Second, GIS permits the combination of different units and levels of analysis into a single dataset through various kinds of table joins. Finally, the location information in GIS can be used to create contiguity and distance matrices for spatial econometric analysis. Taking advantage of these three benefits of GIS, it becomes clear in this study that the long-term emphasis on downtown and entertainment district development has not curbed population emigration from the city and that Housing Conservation Districts may be responsible for higher property vacancy rates.

INTRODUCTION

The use of geographic information systems (GIS) as a tool for program and policy evaluation is beginning to make an impact on the social sciences. This is primarily the result of new developments in GIS programming that permit such systems to run on standard desktop computers and be used by individuals with modest computer and quantitative training. The thematic mapping made possible with software programs like ArcView GIS arms the analyst with the ability to spatially visualize demographic trends within program or municipal boundaries. The display of data in map form is often more useful than tables of numbers to the policy analyst responsible for explaining performance and development results to more qualitatively oriented professionals. Moreover, analysts are finding that the visualization of spatial data, prior to the specification of quantitative models, leads to more accurately specified models. Finally, GIS permits the creation of distance and continuity matrices that can be used to statistically examine spatial clusters and spatial lags in data. This chapter highlights the use of GIS for the visualization of data, model specification, and spatial statistics in a study of neighborhood stability in St. Louis City.

St. Louis City is the ideal illustration of what can be learned through the use of GIS and GIA (geographic information analysis) for three reasons. First, St. Louis City is a separate governmental entity from St. Louis County. St. Louis City is its own county and does not receive any of the benefits of the sub-urbanization of St. Louis County. Second, St. Louis City census geography has stayed relatively constant for twenty years, allowing for a three-period (1970, 1980, and 1990) demographic analysis across both time and space. Finally, the city is loosing population at an extreme rate despite serious redevelopment attempts. The city lost 17 percent of its population between 1970 and 1980, 27 percent between 1980 and 1990, and is projected to loose around 21 percent between 1990 and 2000. In essence, the city will have gone from a population high of 856,796 in 1950 to approximately 300,000 by the turn of the century. This is less than the city's population of 310,864 in 1870. If the downward trend continues, as anticipated, the fiscal survival of the city is doubtful. To analyze this trend, population in neighborhoods is studied.

Numerous variables lead to the increase and decrease of neighborhood residential populations. Some of these variables, such as crime, unemployment, quality of housing stock, and the location of public housing, urban programs can attempt to impact. The degree to which these programs are successful can, in part, be determined by examining demographic trends in neighborhoods and program areas over time and in relationship to other variables. This study examines the effect the city's long-term policy emphasis on downtown business and entertainment district development has on residential neighborhoods. It also shows how target specific projects such as the Housing Conservation Districts have impacted population migration.

DATA COLLECTION AND ANALYSIS

Tools

The substantive focus of this study is on neighborhood residential stability across time and space in St. Louis City. Special tools are needed to conduct this type of spatial research. The desktop version of ArcView 3.1 is the GIS program of choice for this study. Much of the demographic data analyzed came from the Wessex Corporation, which also provides geographic boundaries and street coverage (from census TIGER files) required for geocoding the crime and housing point data.

Also needed is a program that will construct spatial weight/binary contiguity matrices and perform spatial statistics. SpaceStat by Luc Anselin of the Regional Research Institute at West Virginia University is one of the few programs on the market written especially for the statistical analysis of spatial data (Anselin and Hudak 1992). SpaceStat constructs weight matrices and spatial lags. Moreover, SpaceStat comes with a utility that converts ArcView shape files into a format SpaceStat can use to create sparse weight matrices.

Finally, a data conversion program is needed to transfer datasets into numerous different formats. StatTransfer, for example, is used to convert the dataset used with the locational information into the Gauss format required for SpaceStat. It is also used to transform data into a dbase format required by ArcView, and into a Stata format required by the statistical software package Stata. Although SpaceStat performs some non-spatial statistical analysis, it is not a comprehensive statistical package; therefore, the traditional statistical software package Stata is used in addition.

Data

The data used in this study are primarily St. Louis City census block group data. Additional data came from the St. Louis City school board (the number of public school children by race and education level) and the St. Louis City Housing Authority. Individual crime level data came from the St. Louis Department of Police as part of a research project supported by NSF and NIJ grants. St. Louis City, and hence this study, is unique in that city census geography can be held constant for a 20-year period, which is not necessarily true for cities experiencing increases in population and economic development. Municipalities growing in population and area (typically through annexations) also experience growth in the frequency of census geographic units, making spatial comparisons from one decade to the next difficult.

For this particular study, variables are at the block group level instead of the larger tract or voting precinct levels. The block group level is used since it is currently the smallest grid for which these type of census data are available. Also small grid geography best approximates the perceived boundaries of neighborhoods. St. Louis block groups on average contain approximately 800 to 1000 people, while St. Louis census tracts contain 4,000 to 5,000 people. Small grid geography also provides a rich analysis of neighborhood dynamics and offers a helpful contrast to previous research conducted at the larger tract unit. The notion of community and neighborhood is lost at the tract level, as is the sensitivity to change.

The specific variable used to measure neighborhood stability is percent of population age 5 and over living in the same house for five or more years. This measure is taken for 1970, 1980, and 1990, therefore offering measurements for 1965-1970, 1975-1980, and 1985-1990. These variables, given the shortened name 'percent same house', are used to proxy the notion of neighborhood stability, which is little residential turnover. Other key aggregate level variables used to explain neighborhood instability include percent African American population; percent housing units that are vacant; percent population in managerial, technical, or professional positions of employment; frequency of public housing vouchers in a neighborhood; and the presence of public housing units. Immobility and stability are generally reflected by percent housing units that are owner-occupied, percent population age 65 and over, percent families with children 18 years of age or younger, percent unemployed, and percent population at or below the poverty line.

Individual level data for the frequency of aggravated assaults occurring in the city between 1980-1990 is also examined, although a variable for assaults occurring between 1985-1990 is employed in the final model. Homicide data for this same time period were also considered and found not to impact the measure of neighborhood stability because of its low frequency across block groups. Homicide, despite its news-worthiness, remains a rare event. Also, the percent of children attending public schools in a neighborhood are examined but not included in the final model.

Method

Geographic Time Series

Since block group boundaries in St. Louis City have essentially remained the same since 1970, data for 1970 and 1980 can be displayed in 1990 geography, maintaining a constant twenty-year unit of analysis. To transfer 1970 and 1980 data into 1990 geography, the centroids from 1970 and 1980 block groups were spatially joined to the 1990 block group polygons. The spatial join procedure is not error free. There are a few locations where 1990 block groups lack a previous year's centroid, where 1990 block groups are spotted with more than one centroid, and where 1990 block groups have no population. To correct for much of this error, block groups with missing or little data and block groups containing duplicate information were combined with neighboring block groups with similar demographic characteristics.

The elimination of missing data is essential for the construction of the next neighbor matrices used to develop spatial lags. Elimination of missing data from this dataset via block group combinations also necessitates the combination of block group boundaries in the GIS. Hence, analysis is performed on a subset of 546 block groups instead of the total 587. Using boundaries from the Wessex Corporation, census data aggregated by block group are thematically mapped in ArcView GIS prior to model specification. Aggregate data are displayed as polygon groupings (e.g. quantiles or natural break classifications) and individual level data such as aggravated assaults are geocoded (address matched) to geography so as to be displayed as points.

Spatial Lags and Correlograms

One way to get a sense of spatial structure is to construct spatial correlograms of the variables of interest. A spatial correlogram is a pictorial representation of sequential correlations of spatial lags of a single variable with the original single variable (Anselin 1988, Anselin 1992, Kohfeld and Sprague 1994). A spatial lag is a lag in space or a moving-outward in space in all directions to consider the influence of adjacent neighbors on aggregate behavior in a given block group. A 1st order spatial lag includes all the block groups touching the initial block group (e.g. going only one space out in any direction like the King configuration in Chess). A

2nd order spatial lag includes all the block groups touching the 1st neighbors of the initial block group removing all connections that were included in the first neighbor matrix, and so on. This makes each matrix disjunct with every other matrix and therefore preserves their independence.

To obtain spatial lags for the next neighbor structure, contiguity matrices are needed. For St. Louis City block groups, contiguity matrices can be constructed by exporting GIS boundary files into the software program SpaceStat. SpaceStat contains a utility that converts ArcView GIS boundary files (containing location information) to the file format BND. The BND or boundary format is used in SpaceStat to construct a binary contiguity matrix in sparse form (a GAL file format). Next, higher order contiguity matrices are made from this original matrix via SpaceStat's weight transformation function. By order 13 some of the neighbors are no longer connected - meaning there are instances of no-neighbors. Consequently, spatial statistics are only calculated on higher order contiguity matrices through order 12.

Twelve spatial lags are then constructed for key variables from these matrices. The spatial correlograms presented in this chapter are for percent same house (1970, 1980, and 1990). The percent same house variables are transferred via StatTransfer into a Gauss dataset, which can be read by SpaceStat. A first order spatial lag is then constructed from the first neighbor matrix, a second order lag from the second neighbor matrix, and so on. Next, the spatial lags are correlated with the original variables, and the first column of the correlation matrix is input into a new dataset from which the spatial correlograms are made.

Thematic Mapping and Spatial Autocorrelation

In addition to examining the spatial structure of stability through spatial correlograms, neighborhood stability or instability can be seen in time and space through thematic mapping. Here, thematic mapping is considered a method of both visual analysis and model specification. Mapped variables advance the understanding of how multiple factors affect stability, and this knowledge can then be used to choose and configure model variables.

Model residuals can even be spatially correlated and mapped for the inspection of spatial dependence. It is important to examine the spatial structure of the residuals for spatial autocorrelation, which can affect model coefficients (Anselin 1988, Dubin 1992, Kelejian and Robinson 1992). Constructing the spatial correlograms from the residuals of different models can assist in evaluating model specification and provide evidence of the success of the models in removing the inherent spatial dependence found in the original variables. The process used in this study is to save model residuals from each different model using a standard statistical package such as Stata, and transfer the residuals into the required Gauss format for use in SpaceStat to construct the spatial lags. Once the spatial lags are constructed, they are re-transferred back into Stata, where matrices and correlograms are constructed.

The correlograms of model residuals allow visualization of possible spatial autocorrelation problems; however, they are not designed to pinpoint local clustering. Although spatial dependence can be minimal for the model as a whole, there may be small clusters of block groups where the model still over or under predicts. To visually test for local clustering, model residuals are recoded into three groups:

- an over predicted group which consists of residuals one standard deviation or more above the residual mean
- an under predicted group which consists of residuals one standard deviation or more below the residual mean
- the more accurately predicted group which consists of residuals within one standard deviation above or below the residual mean.

The grouped residuals are then displayed in a thematic map. A good scheme for polygon coloring can easily highlight areas of local clustering.

FINDINGS

Spatial Correlograms

One way to systematically explore the spatial structure of neighborhood stability is to actually examine the extent to which stability in one block group is related to stability in surrounding block groups. Spatial independence is indicated by a randomly occurring distribution of persons living in the same house in a block group throughout the city. The values of the observations for the same-house variable are not expected to be independent of their location; neighborhood context most likely influences moving behavior even at the aggregate level.

Figure 1 shows correlograms for people living in the same house by block group across years. The high correlation from year to year indicates that stability is highly spatially structured, otherwise the correlogram bars would be randomly distributed between negative and positive correlation coefficient values, and those correlation values would be low. There is positive correlation between people living in the same house by block groups to order 6 (or six block group neighbors out) in 1970 and 1990, and order 5 in 1980, before the correlation coefficient becomes negative. However, there is slight variation in the magnitude of correlation across years. In 1970 and 1990, the correlation coefficient for the correlation of persons living in the same house with persons living in the same house in contiguous block groups starts above 0.4 (about 0.44 in 1990 and 0.53 in 1970), while the correlation coefficient starts below 0.3 in 1980.

In short, this spatial correlogram can be interpreted as indicating a divergence in 1980 from established spatial structures of neighborhood stability. This divergence corresponds to the pro-gentrification and urban renewal policies of the

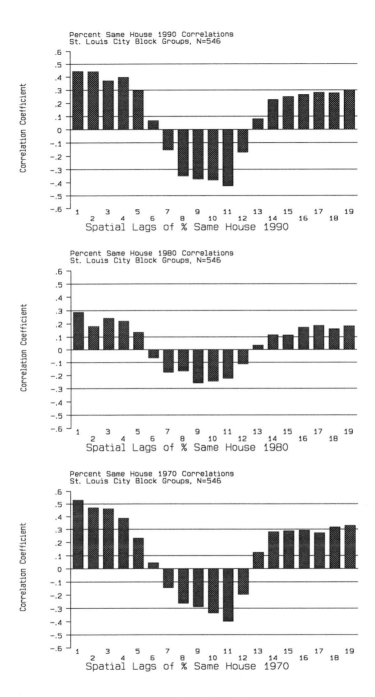

Figure 1 Spatial correlograms

221

1980s, which produced pockets of high-end residential and commercial development in older and relatively low-income city neighborhoods.

Thematic Maps

The spatial structure of stability is also visible in Figure 2. These thematic maps for 1970 and 1990 are similar, with stability exhibiting different patterns in 1980. During the 1980s, city planners did a great deal to encourage migration back to the city from the outline county areas. Programs such as the City Living program, generous tax abatement, and other redevelopment projects led to a period of gentrification not present in the 1990s. The effects of such programs and trends can be seen in both in the spatial correlograms and the thematic maps. Most notable is the pattern of higher stability in the north section of St. Louis in 1980 as compared with that same section in 1970 and 1990.

St. Louis City led the U.S. in white flight during the 1970s and is still one of the most residentially segregated cities in the nation. Racial tension is considered an important factor in segregation as well as population instability. The migration of African Americans from north St. Louis to south St. Louis corresponds to the flight of predominately white middle class population to the suburbs. The pattern of African American population growth and migration is visible in Figure 3.

Besides race relations, two commonly cited pushes in population movement are the location of crime and public housing, with public housing being generally associated with crime, poverty, and neighborhood deterioration. Figure 4 depicts the occurrence of aggravated assaults in St. Louis City between 1985 and 1990 and the percent of the population age 5 and older living in the same house between 1985 and 1990 (the variable used to indicate stability). Using quantile classifications, the white (clear) areas of the maps are areas for which there is high stability (64 percent - 100 percent living in the same house since 1985 or earlier) and low crime (few incidents of assaults between 1985 and 1990). The most notable white areas on either map are in the southwest section of the city, where there is high population stability and low incidence of aggravated assaults. Although the white area is visible in the south section of the city in both maps, a notable scatter of white is visible in the north section of the city for percent in the same house but not for assaults. One explanation is that for the population remaining in the city in 1985, either because of immobility or choice, factors other than crime take precedence in their moving decisions.

As expected, the location of public housing, both public housing complexes and housing voucher recipients, roughly mirrors the location of aggravated assaults. Figure 5 shows the location of public housing complexes that are non-elderly and contain 50 or more units. Figure 5 also displays the location of recipients of public housing vouchers. Comparing Figure 4 with 5, it is clear that where there are fewer aggravated assaults there are fewer voucher recipients and no public housing complexes. The concentration of public housing complexes in the east central section of the city indicates that the overwhelming site choice for large non-elderly public housing complexes is near the river front industrial district

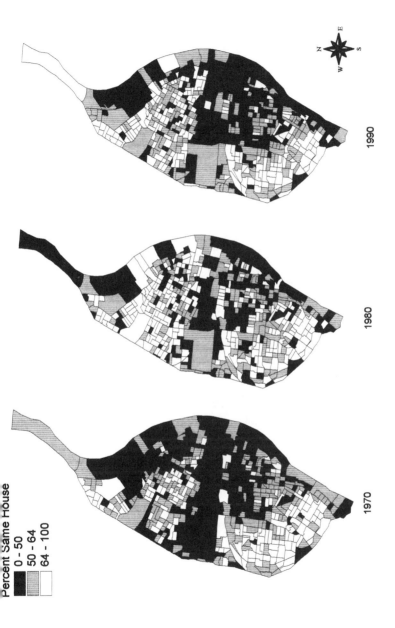

Figure 2 Percent same house 1970-1990, St. Louis City block groups, N = 546

223

Figure 3 Percent African American population 1970-1990, St. Louis City block groups, N = 546

Figure 4   Percent same house 1990 and aggravated assaults 1985–1990, St. Louis City block groups, N = 546

Aggravated Assaults 1985–1990
N = 33,101

0 - 50
50 - 64
64 - 100

N
W   E
S

Percent Same House
- 0 - 50
- 50 - 64
- 64 - 100

Non-Elderly Public Housing Complexes
Where Units = > 50, N = 12

Housing Voucher Recipients, N = 6,969

N
W E
S

Figure 5. Percent same house 1990 and public housing 1996, St. Louis City block groups, N = 546 (east city boundary) where there are few residences. In fact, a windshield survey of these complexes indicates that they are largely the only development in their census block group.

Again, in the predominantly white south side of St. Louis city, the most stable neighborhoods are those in which there are few housing voucher recipients and no public housing complexes, while the predominantly black north side contains stable neighborhoods characterized by aggravated assaults and voucher recipients. It may be that those residing in northern neighborhood are either unable to move or less sensitive to crime and public housing, or are public housing residents themselves. Based on the patterns exhibited in these maps, a decision is made to include variables for crime and public housing in the specification of spatial econometric models. Also, given the difference in the visual correlation between assaults and stability, and between housing vouchers and stability, the econometric model is specified for each section of the city rather than the city as a whole.

Results of Spatial Econometric Models

Importance of Homeownership

Given the information gained from exploratory research via the visual inspection of GIS maps and spatial correlograms (not all are presented in this chapter), coupled with previous knowledge of neighborhood change in St. Louis City, regression models can be specified with some degree of accuracy. To assess variation in the causes of stability for different city neighborhoods and hence different city racial groups, a spatial econometric model is specified for the predominantly black north side, the racially mixed central corridor, and the predominantly white south side. The results are presented in Table 1. The single variable that is statistically significant across racial categories is percent owner-occupied housing units. The positive coefficient and statistical significance of the coefficient for percent owner-occupied housing units suggests that policy programs designed to increase home ownership can be applied city-wide, across racial groups, and be effective.

Differential Importance of Other Model Variables

A few other variables reach significance in the models qualified by race. Percent of the population in the same house in 1980 is significant for racially mixed (mostly central corridor) and predominantly white (mainly south side) sections of the city, but not for predominantly black block groups. This result marks the notable demographic change and out-migration that has occurred on the north side. This change is strong enough to reduce the ability of stability in 1980 to predict stability in 1990. Age acts as a stabilizing factor in predominantly white and racially mixed neighborhoods; the positive and significant relationship does not carry over into predominantly African American neighborhoods most likely

Table 1   Regression models for percent of population living in same house for 5 or more years (between 1985 and 1990) in St. Louis City by racial groups

Variable	Coefficient (t-value in parentheses)		
	White	Mixed	Black
Percent Same House 1980	0.247 (4.173)***	0.130 (2.293)**	0.189 (0.314)
1st Order Spatial Lag of Percent Same House 1990	-0.036 (-0.241)	0.057 (0.336)	-0.066 (-0.355)
Percent of Housing Units Owner Occupied 1990	0.281 (5.977)***	0.342 (4.658)***	0.258 (3.559)***
Percent of Population Age 65 and Over 1990	0.294 (2.624)***	0.423 (3.445)***	0.233 (1.300)
Percent of Families with Children 1980	0.1223 (1.743)*	0.020 (0.287)	-0.038 (-0.531)
Percent Unemployed 1990	0.076 (0.493)	0.056 (0.413)	0.010 (0.096)
Percent of Population at or below the Poverty Line 1990	0.258 (0.225)	0.098 (1.093)	-0.080 (-0.975)
Percent of Housing Units Vacant 1980	-0.196 (-0.878)	-0.091 (-0.979)	0.228 (1.896)*
Percent of Population in Managerial, Technical, or Professional Positions 1990	-0.049 (-0.723)	-090 (-1.189)	0.051 (0.517)
Aggravated Assaults per 100 People 1985-1990	0.178 (0.524)	-0.803 (-2.853)***	0.0878 (0.136)
1st Order Spatial Lag of Aggravated Assaults 1985-1990	-1.965 (-0.892)	0.492 (0.705)	0.839 (1.085)
Housing Vouchers per 100 Households through 1996	-0.672 (-1.767*)	0.141 (1.059)	-0.229 (-1.226)
Interaction of Unemployment 1990 and Location of Public Housing Complexes	3.750 (0.562)	0.118 (0.483)	0.485 (2.838)***
Location of Public Housing Complexes (Dummy)	-8.373 (-0.837)	-4.111 (-0.997)	-14.078 (-3.153)***
Constant	22.261 (2.229)**	18.789 (2.009)**	50.417 (3.564)***
R^2	0.58	0.40	0.23
Adj. R^2	0.54	0.34	0.16
F (14, 183)	17.73	6.81	3.62
N	198	160	188

Source: 1980 and 1990 U.S. Bureau of the Census; St. Louis Department of Police; St. Louis City Housing Authority.

*** $p < .01$, ** $p < .05$, * $p < .10$

because the infrequency of clusters of this age group. Aggravated assaults are a negative influence on stability in the racially mixed neighborhoods probably because racially mixed areas are less stable (more susceptible to neighborhood tipping when income is also mixed) than non-racially mixed neighborhoods. Aggravated assaults (the coefficient being negative and significant) could fuel the tipping dynamic.

Also, a negative and significant coefficient for rate of housing voucher recipients in predominantly white neighborhoods suggests sensitivity to the presence of voucher recipients and the demographic change that the increase in voucher housing represents. In white neighborhoods more so than non-white or mixed neighborhoods, housing vouchers encourage out-migration. The positive and significant coefficient on the variable representing the interaction between unemployment and location of public housing in predominantly black neighborhoods supports the previously discussed idea that such interaction renders people immobile in their current environments, more so than unemployment alone. In all three areas, the coefficient for the presence of public housing complexes is negative; it most likely only reaches significance in the predominantly African American block groups because of the frequency of public housing, especially when compared with the south side.

The compelling inference to be drawn from these results is that strategies to promote homeownership can be applied citywide with confidence that the result will be increased stability. These results also indicate that the total abandonment of central city residential districts is unlikely. Unfortunately, the groups of people least likely to relocate are those with limited financial resources, and hence limited mobility. Although the city retains population in this circumstance, it will undoubtedly experience fiscal crisis, with the tax base being unable to support the service needs of the population. In this light, policy programs encouraging homeownership should make special efforts to target middle-income populations.

Elimination of Spatial Dependence

To ensure that the model estimates are as accurate as possible, it is necessary for any spatial autocorrelation to be modeled out of existence. Evidence of spatial dependence in the model can be found in spatial correlograms of the spatial lags of model residuals at various stages of model development. Correlogram 1 in Figure 6 presents results from the model with both types of variables – those intended to predict stability and those to predict instability. Here, the spatial structure is attenuated, starting at correlation coefficient 0.3 with the pattern of positive correlation lasting only through the third lag. With the inclusion of the owner-occupancy variable in the model (correlogram 2 in Figure 6), spatial autocorrelation is reduced even further with correlation coefficients never exceeding 0.1, but the spatial patterning still exists. The residuals for the full model as presented in the third correlogram in Figure 6 indicate the elimination of spatial dependence, evidenced by the loss of trend in the lags.

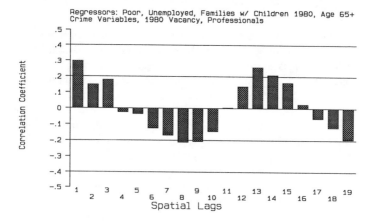

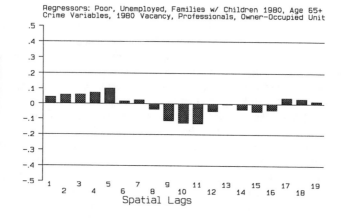

Figure 6   Correlated spatial lags of model residuals

Although the correlogram of the residuals for the full regression model indicates the elimination of spatial dependence in the model, there still may be pockets of local clustering where the model over or under predicts. Therefore, model residuals are mapped in ArcView. The light gray polygons in Figure 7 indicate block groups for which percent same house 1990 is over predicted (more than one standard deviation above the mean) and the black polygons indicate block groups for which percent same house 1990 is under predicted. This map suggests that, overall, the full model tends to be conservative, under predicting more than over predicting neighborhood stability. Moreover, what is gained here and lost in spatial statistics is an understanding of the regions in which clustering of over or under prediction occurs. For instance, under prediction of neighborhood stability is clear in each of these illustrations of the city for the central corridor, the only region of the city that is racially integrated. Under prediction is also evident along the city's eastern boundary (the Mississippi River front) where residential populations are small.

Linking Residential Mobility and Downtown Redevelopment

One very general but overwhelming conclusion can be drawn from these results. The city has emphasized large, visible redevelopment projects, such as the restoration of the central business district, the construction of new sports stadiums (the TWA dome and the new Kiel Arena), and the introduction of riverboat gambling. But these projects have done little to keep people from moving out of the city and from attempting to escape urban problems within city boundaries. Continued out migration of city population has produced fiscal stress for city government, and will undoubtedly result in a situation of fiscal crisis in the near future. Instability within city neighborhoods can produce an environment conducive to city abandonment. The continued mobility from north to south St. Louis makes delivery of city services and maintenance of urban infrastructure difficult. The inability of city administration to properly meet the service needs of its population and maintain infrastructure in areas of high population migration may further heighten moving behavior, thus exacerbating the problem. However, the results also suggest that programs focusing on homeownership may better serve the city in its effort to retain population than continued concentration on sports, entertainment, and business district redevelopment.

REDEVELOPMENT POLICIES

St. Louis planners recognize the need to redevelopment or at least maintain the quality of the city housing stock in order to entice new buyers and to keep residents from moving elsewhere. This is a particular challenge since most of the city's housing was built before 1940. One program solution implemented in the city is the Housing Conservation District (HCD). The stated goal of the HCD ordinance is to preserve the quality of housing and protect citizens and

Figure 7 Mapping residential groups from full regression model, St. Louis
City block groups, N = 546 (Y = percent same house 1990)

neighborhoods from housing deterioration that threatens health, safety, and welfare. All residential dwellings (rental and owner-occupied) within a HCD are to be inspected by an HCD building inspector upon change of occupancy. The Union Electric and Laclede Gas utilities inform the city building division of an occupancy change. The property owner must obtain an occupancy permit (Certificate of Inspection) for the occupant within 37 days of occupancy change (7 days to apply and 30 days for issuance). If property improvements are needed, the owner has 60 days to make repairs. Occupancy permits are limited to a cost of $40 per permit. Inspections mainly cover health and safety concerns. Interior inspections include:

- wiring, plumbing, mechanical equipment (heating systems)
- trash, sanitation, and rodent/insect infestation; water damage and chipping plaster or tile
- ability of doors to properly close and lock, and the presence of smoke detectors
- structural integrity of the foundation, bearing walls, and floors
- minimum gross floor area, sleeping room area, living/dinning/kitchen area per occupant
- special requirements for basement occupancies.

Exterior inspections include chipping paint, broken glass, the condition/existence of handrails, and overall condition of front and back porches. In general, the HCD ordinance requires stricter enforcement of building code standards than in other areas of the city. Failure to obtain an occupancy permit or to correct cited housing problems can result in a fine of up $500 and/or 90 days imprisonment. HCD are established for a contiguous area containing at least 500 dwelling units by application to the Building Commissioner. This application must be approved by the Community Development Agency (CDA) in the form of a Statement of Feasibility, and passed into city ordinance by the Board of Aldermen. An alderman or group of aldermen can propose a HCD for parts of different wards or an entire ward. Thus far, HCD do not strictly follow ward boundaries.

Unfortunately, HCD are thought to have the unintended consequence of increasing property vacancy rather than improving property maintenance. Some landlords, instead of bringing property up to the standards set by the HCD ordinance, simply abandoned it, leaving the property to be owned by the city. In 1950, the city had enough housing to accommodate almost a million people. In 1990, there was less than 400,000 people left in the city. Although, some of the unwanted housing had been demolished, much of the housing still remains, creating a circumstance where supply exceeds demand. With the sale of housing being difficult in many areas of the city, owners often choose to abandon properties rather than pay property tax and incur costly maintenance bills. In April of 1997, the city's land redevelopment authority (LRA) owned almost 10,000 scattered buildings, 6,984 of which had been vacated. If the city cannot find a buyer, it eventually boards the property up. Many of these properties incur further decay in until the city demolishes it years later.

Figure 8 illustrates St. Louis City (HCD) and the percent of vacant housing units in 1980 and 1990, with the advent of HCD in the 1980s. Clearly, the region within the HCD in southeast St. Louis City has experienced an increase in housing vacancy, moving from less than 10 percent vacant, to between 10 and 20 percent vacant. This increase is also evident in the mid-north HCD and the northwest HCD. To be certain of the connection between vacancy rates and HCD, a longer time series is being obtained.

CONCLUSIONS

This chapter illustrated the use of GIS in the examination of neighborhood instability patterns in St. Louis City. GIS was also used to evaluate the city's Housing Conservation District program. The GIS software ArcView permitted thematic mapping which illustrated demographic changes in city census blocks in time and space. The thematic information displayed in these maps was then used to specify spatial econometric models while the location information from the maps was used to produce contiguity matrices necessary to make spatial correlograms.

The spatial correlograms suggest that the redevelopment policies in place in the 1980s may have advanced neighborhood stabilization more so than any policy in place during the 1970s or 1990s. The spatial correlograms also indicate that the model coefficients are not being affected by spatial autocorrelation. The models themselves suggest that the key to stabilizing neighborhoods is increasing homeownership. Current St. Louis City redevelopment policies focus mainly on the redevelopment of the city's entertainment and business districts. One exception is the Housing Conservation District program which attempts to increase the maintenance of the city's housing stock. Unfortunately, a thematic map of vacancy rates and HCD suggests that HCD may be having the unintended consequence of increasing housing vacancy.

ACKNOWLEDGMENTS

The author would like to thank Carol Kohfeld, John Sprague, and Lana Stein for their guidance in this study.

REFERENCES

Anselin, L. (1992). "Space and applied econometrics." *Regional Science and Urban Economics, 22, 307-316.*

Anselin, L. and Hudak, S. (1992). "Spatial econometrics in practice: a review of software options." *Regional Science and Urban Economics, 22, 509-536.*

Anselin, L. (1988). *Spatial econometrics: methods and models.* Kluwer Academic, Dordrecht, The Netherlands.

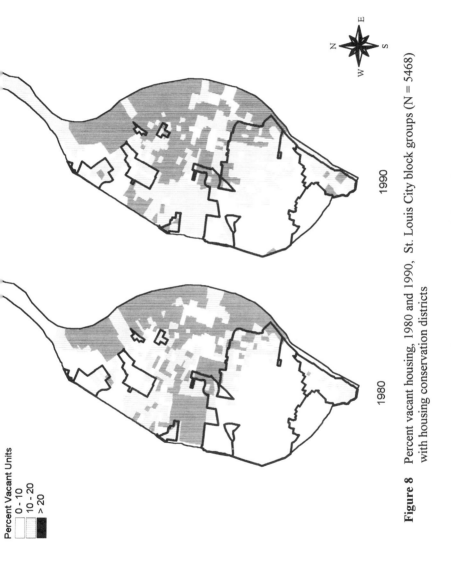

Figure 8 Percent vacant housing, 1980 and 1990, St. Louis City block groups (N = 5468) with housing conservation districts

Dubin, R.A. (1992). "Spatial autocorrelation and neighborhood quality." *Regional Science and Urban Economics, 22, 433-452.*

Kelejian, H.H. and Robinson, D.P. (1992). "Spatial autocorrelation: a new computationally simple test with an application to per capita county policy expenditures." *Regional Science and Urban Economics,* 22, 317-331.

Kohfeld, C.W. and Sprague, J. (1994). "Exploiting spatial autocorrelation as a tool in model specification." Political Science Paper Number 250, Washington University in St. Louis, *Annual Meeting of the American Criminological Society*, November 8-13, Miami, FL.

CHAPTER 14

Civil Engineering Education

Said Easa, Songnian Li, and Yaoyu Shi

Integrating geographic information system (GIS) concepts into civil engineering (CE) education is very important not only to meet the urgent needs of non-GIS professionals in engineering, but also to teach students relevant skills in spatial analysis, reasoning, and data processing. Recent improvements in computer hardware and software also make it possible for GIS technology to move effectively into the education arena. This chapter reviews current status of GIS uses and developments in civil engineering education and presents guidelines regarding GIS course setting, educational methods, and infrastructure needs. To determine current uses of GIS in education, a questionnaire survey of Canadian universities was conducted. The results show that 53% of the universities that responded have GIS courses (or CE courses in which GIS concepts are included) and 30% plan to include GIS contents in their future curriculum. GIS course settings at three educational levels are presented. Educational methods of teaching GIS in the classroom include demonstration teaching, hands-on interactive learning, and self-pace learning. Internet sources for teaching GIS are also presented. GIS infrastructure needs in CE education include software and hardware, data sources, and laboratory setting.

INTRODUCTION

GIS means more to education than just producing maps and other spatial information. It can affect the whole educational experience of students and teachers and meet the urgent application needs of GIS professionals (ESRI 1995a). GIS applications are limited by such factors as high system cost, lack of digital databases, and lack of experienced personnel (Horak 1994). Also, Lourie (1994) explained that GIS education has become urgent because of the lack of qualified specialists with the ability to create and use GIS in academic institutions, companies, and organizations. People are one of the four major components that are important in GIS development, besides hardware, software, and data. Most important, there exists an urgent need to have professionals knowledgeable in GIS technology and its implementation in non-GIS specialities (such as civil engineering) because GIS is another tool that can be used in the civil engineering (CE) analysis, design, and decision making.

Reprinted with permission from American Society of Civil Engineers: "GIS technology for civil engineering education. S. Easa, S. Li, and Y. Shi, *Journal of Professional Issues in Engineering and Practice* (1998), 124(2), 40-47 (slightly revised)

Recent improvements in computer hardware and software allow GIS technology to move effectively and affordably into the educational arena. With high-performance and low-cost desktop computers, many desktop-version GIS software have evolved from their mainframe counterparts or have been newly developed, which could accommodate most GIS application and educational requirements. The Internet, World Wide Web (WWW), and client/server technologies make digital databases exchangeable on-line and supplement the need for spatial data and relevant metadata. During the next decade, GIS will prove to be a popular educational tool in relevant disciplines, like computer education was in the 1960's.

GIS education can be offered at different levels, including pre-university education (elementary and high schools), high learning education (colleges and universities), and professional education (institutes and organizations). At present, GIS is mostly developed at the university/college level as courses (or new programs) at the undergraduate/graduate level in geoscience-related disciplines such as geography, geomatics, geology, forestry, and surveying. The developments mainly focus on educating GIS specialists who will be able to design, build, analyze, and maintain the systems, not on educating GIS professionals. GIS education in civil engineering at universities is limited. The discipline of civil engineering includes a wide range of specializations that are related to spatial-data processing. These specializations include geotechnical engineering, project management, highway design, municipal engineering, transportation engineering, urban planning, wastewater and waste management, environmental engineering, landform assessment, re-engineering, and utilities management. Integrating GIS into CE education is very important not only to meet the urgent needs of non-GIS professionals in engineering, but also to teach students relevant skills in spatial analysis, reasoning, and data processing.

This paper first reviews current GIS uses and developments in CE education at the university level, including the results from a survey of Canadian universities conducted in this study, followed by suggestions for GIS course setting. Educational methods and infrastructure needs in CE education are then presented.

CURRENT STATUS

GIS educational status in civil engineering includes incorporating new technologies into existing curriculum and using GIS as a computer-aided instruction (CAI) tool to teach geographic data-related courses. Table 1 lists the GIS and GIS-related courses that are currently available at various academic institutions. These courses have been identified through a literature review (library and Internet) and a survey of Canadian universities.

Literature Review

The library literature review was confined to publicly published journals and magazines, and university calendars. Only a limited number of journal and magazine

Table 1   GIS courses available at civil and environmental engineering departments
          in Canada, U.S.A, New Zealand, and Switzerland

University name	Course name	Type[a]	Level[b]
Colorado State Univ.	GIS; Advanced GIS	I	UG
Kansas State Univ.	GIS (as elective course)	I	G
Iowa State University	Analytical Photogrammetry GIS	I	UG
New Jersey Inst. of Tech.	Geographic Information Systems	I	UG
Oregon Inst. of Tech.	Geographic Information Systems	I	UG
Penn State Univ.	Civil Engineering Measurements	II	UG
Queens University	Surveying Field School (spring)	II	UG
Royal Inst. of Tech.	GIS for Civil Engineers	I	UG
Roy. Mil. Coll. of Canada	Geomatics	II	UG
Ryerson Polytech. Univ.	Introduction to Geomatics	II	UG
State Univ. of New York	Geo., Surv., CAD, GPS, and GIS	II	UG
	GIS Applications	I	G
The Ohio State Univ.	Intro. to GIS; Des. & Impl. of GIS	I	G
Universite de Moncton	Geomatics	II	UG
Univ. of Alberta	Surveying	II	UG
	Advanced Topics on GIS	I	G
Univ. of B.C.	Plane Surveying	II	UG
Univ. of Calgary	Remote Sensing in Hydrology	II	G
Univ. of Canterbury	Geographic Information Systems	I	UG
Univ. of Colo. (Denver)	Geographic Information Systems	I	G
Univ. of Guelph	GIS in Environmental Engineering	I	UG
Univ. of Manitoba	Transportation Engineering	II	UG
Univ. of New Brunswick	Geographic Information Systems	I	UG
	Advanced Surv.; Eng. Hydrology	II	UG
Univ. of Oklahoma	GIS Applications in CE Science	II	UG
Univ. of Ottawa	Measurements I	II	UG
Univ. of Regina	Surv., Mapping and Info. Systems	II	UG
Univ. of Saskatchewan	Transportation Engineering I	II	UG
Univ. of Wis.-Madison	Principles of Land Info. Systems	I	UG

[a] I = GIS course; II = course containing GIS concepts.
[b] UG = undergraduate course; G = graduate course.

papers related to GIS education in civil engineering and GIS education in general,
were found. Some universities have already developed GIS or GIS-related courses in
the curriculum of civil and environmental engineering departments. These courses
were found at both the undergraduate and graduate levels from university calendars.

The Internet literature review was conducted for universities that have CE
programs, on-line conference proceedings, and other relevant papers. In order to
obtain more information about current developments, nearly all Internet homepages
of CE departments in New Zealand, Australia, and Canada, and many universities in
United States, United Kingdom, and China were searched. It is noted that some

universities do not put their course information on the Internet, and therefore the search in this case was in vain. The literature found on the Internet was mainly related to GIS education in general. Several aspects related to GIS education in CE need to be addressed, including implementation issues, re-engineering issues, and human factors.

Survey of Canadian Universities

To find the current status of GIS in CE education at Canadian universities, a survey was conducted using questionnaires. The questionnaire forms were sent to 34 universities that have civil and environmental engineering departments. The list of these universities was obtained from the Internet (http://www.civeng.ca/chairs/). The questionnaire included questions about current GIS courses, GIS concepts in other related courses, and future GIS courses (Figure 1). The number of departments that responded to the questionnaires was 23, representing a response rate of 68%.

The results of the survey are shown in Table 2. Among the 23 responses, only two departments have GIS courses at the undergraduate level, ten departments have GIS-related courses (CE courses that include some GIS concepts) mostly at the undergraduate level, and seven departments expressed a desire to include GIS contents in future curricula, especially integrating GIS concepts into CE courses. Examples of GIS-related courses are surveying, remote sensing, geomatics, hydrology, and transportation engineering. Future curriculum changes will depend on the availability and research interests of faculty members, and the understanding of the GIS role in CE education. Some departments also consider offering GIS courses through other academic departments such as the Department of Geography and the School of Information &Technology.

Current Developments

At present there are two categories for introducing GIS concepts into CE curriculum in high-learning institutions (Table 1): (1) developing GIS courses, such as GIS, advanced GIS, and Spatial Information Systems and (2) integrating foundations of GIS into other related courses, such as surveying, photogrammetry, and geomatics (Lourie 1994). Most of these courses are taught at the undergraduate level with only a few at the graduate level. The GIS concepts included in these courses vary between the preceding two categories. Within the first category, the concepts mainly include topics relevant to spatial-data management, spatial analysis, application modeling, engineering applications, and introduction to hardware and software. The major purpose of the second category, however, is to teach students introductory knowledge of GIS and, to some extent, engineering applications.

The integration of GIS education into CE curriculum depends on such factors as accreditation standards, research interests of faculty members, educational budgets, and need for curriculum innovation. For example, the National Standards for Applied Science and Engineering Technologists in Canada is a publication that sets out

Table 2   Results of survey of Canadian universities

Types	Grad. level	Undergrad. level	Both levels	Total no.	% of responses
GIS course	0	2	0	2	9%
GIS-related course	2	6	2	10	44%
Future GIS or GIS-related course	1	5	1	7	30%

GIS-COURSE QUESTIONNAIRE

University Name: _____ Respondent Name: _____

1. Do you have any GIS courses in your curriculum?

Yes ____ No ____ If yes, please specify the following items:

Course Name   Main Content Description

a. _____

b. _____

2. Do you have any civil engineering courses that include some GIS concepts in your curriculum?

Yes ____ No ____ If yes, please specify the following items:

Course Name   GIS-related Content Description

a. _____

b. _____

3. Do you plan to include GIS courses into your curriculum in the near future?

Yes ____ No ____ If yes, please specify the following items:

Course Name   Planned Contents

a. _____

b. _____

Figure 1   GIS-course questionnaire used in the survey

standards for technologists in applied science and engineering disciplines (CNSASET 1994). It states that "A civil engineering technologist is able to apply concepts of geographic information systems (GIS) and Global Positioning Systems (GPS) to civil works for the surveying area of competence." This kind of standards may have resulted in some non-GIS courses having GIS concepts in their syllabi.

According to our survey, several CE departments state that including GIS concepts in their curriculum will depend on faculty members' interests. Currently, the lack of human resources for teaching GIS at universities is still one of the major problems. Other departments state that it is difficult for them to afford the lab facilities given the present budget constraints. A few departments have allowed their students to select courses from other relevant departments (such as geography) which have well developed GIS courses. This is a good way to introduce civil engineering students to an introductory or even advanced GIS.

Business incentives are important in developing the GIS technology in CE departments. Vendors can provide financial support for the development of GIS in education for repayment of future use of the vendor's software and software producers can develop low-cost educational GIS software including tutorial versions. Several GIS software vendors and developers have realized the importance of GIS education and have taken actions to support it. Examples are:

- ESRI has provided *University Kit* (including Arc/Info Lab Kit, PC-Arc/Info Lab Kit, ArcCAD Lab Kit, and University Site Licence)
- CARIS has provided *GIS for the Curious* as a special package for a GIS beginner or a person who has some exposed knowledge
- IDRISI has provided its package to university students and faculty at a low cost.

But, at present, the technical and financial support from vendors and developers to high-learning institutions is still inadequate. To some extent, when software companies help universities establish GIS laboratories, they will gain many potential users who will be in charge of future GIS implementations.

GIS COURSE SETTING

GIS is a tool by nature, not a goal in itself. Education appears to deal with concepts and principles of GIS, whereas training seems to deal with acquiring skills for using specific software (Piscedda 1994). For educating civil engineers, both education and training are necessary. This understanding would provide a guideline for integrating GIS into CE curricula. That means that we should focus on the practical, not theoretical, orientation in shaping GIS education in civil engineering without losing the understanding of the basic theory. For example, basic concepts and skills, organizational setting approach and industrial applications should be discussed in GIS courses. Also, learning the basics of the software used in teaching is necessary

Table 3   Proposed GIS courses in civil engineering curriculum

Course name	Level[a]	Lecture (hrs/wk)	Lab (hrs/wk)	Syllabus description
Introduction to GIS	UG	3	2	Basic concepts, including terminology, origin, and history; data acquisition, structure, and management; spatial analysis and modeling; hardware and software; and generic applications
Advanced GIS	UG/G	2	3	Details on GIS architecture and functionality, practical applications with more hands-on experiences, and organizational setting approaches
GIS Seminars	UG/G	2	0	New GIS software, new applications and developments, and specific engineering problem-solving techniques

UG = undergraduate course; G = graduate course.

o help students practice hands-on skills and enhance understanding of GIS.

As such, GIS courses developed in CE curriculum should have three levels (Table 3). The first level is about introduction to GIS: basic concepts and functionality, hardware and software, and basic applications with some hands-on exercises. The goal of this level is to provide students with sufficient background both for using GIS as a CAI tool in learning and for studying advanced GIS courses later. The second level is advanced GIS that includes in-depth discussions about GIS: details on GIS structures and capabilities, spatial modeling and analysis, system setting, and application techniques with emphasis on hands-on practice. This mainly prepares students for future careers that involve GIS applications. The third level is about new developments in GIS. The courses at this level are primarily presented in the form of seminars. Here the students will be exposed more to the recent developments on GIS technology, such as new GIS software, new applications, and GIS solutions related to special engineering problems. This level crosses over from education to training such that practical skills are developed.

The basic requirements of the GIS courses for the preceding levels, including lecture and laboratory hours, are suggested in Table 3. The instructors should modify these requirements according to the specific university curriculum. A detailed GIS course outline at the undergraduate level is presented in Table 4. The description includes course name, academic level, objective, syllabus, contents, and organization. We hope this outline will be useful as a reference for further discussion and implementation.

Instead of developing GIS courses at different levels, some departments may consider including GIS concepts into surveying-related courses. In this case, it is appropriate to include only the basic concepts, GIS principles, and methods of applying specific software. Actually, this is a good practice because surveying-related courses have already provided students with adequate knowledge on measurements, accuracy, maps, and other concepts that are relevant to spatial information. This is also a shortcut to integrating GIS concepts into CE curriculum.

When incorporating GIS courses in CE curriculum at a certain educational level, it is important to define the professional goals of students, key contents, employment needs, and the role of different GIS software. The conflicts between teaching GIS concepts and the employers' needs for using currently available software are always a challenge for GIS educators. The most commonly asked question is what should we teach in academic GIS courses about software: is it basic structure, programming, or users manual? GIS education should not closely depend on specific software (Lourie 1994). With so many software in the market, the purpose of including specific software in a GIS course should be to explain the concepts, structure, and applications of GIS (not those of specific software). The software is only a tool that is used to aid teaching GIS in a university course. However, as the need for in-depth GIS knowledge and training increases in CE, it would be better for a university to partner with a leading software vendor to benefit both students and future employers.

EDUCATIONAL METHODS

GIS education, like education of other modules, can be achieved through theory, exercise, and applications (Piscedda 1994). Because GIS education involves complicated technology, a rapidly changing field of study, and strong application orientation, it seems that workshop and laboratory teaching are especially important for this kind of education (Cremers 1996). The purpose of integrating GIS concepts into CE curriculum is to educate non-GIS professionals to deal with GIS-related applications in civil engineering. Any teaching method must have strong orientation to aid the student's practical abilities instead of just theoretical knowledge.

Generally speaking, there is no best way for teaching GIS technology to the students in the classroom. But there is really a better way to study GIS; namely, using GIS as a CAI tool in teaching students who are already taking GIS courses. This method changes the existing teacher-entered, group-oriented paradigm to a more individual, small learner-centered environment. Using GIS to show real-life practical applications in non-GIS courses can also help students understand the concepts better. Whether using or teaching GIS, the following scenarios can be used in classroom teaching (ESRI 1995b, Chang et al. 1995):

Table 4   Suggested GIS course outline for undergraduate civil engineering

Item	Description
Course Name	Introduction to Geographical Information Systems for Civil Engineering
Academic Level	Undergraduate
Objective	To introduce students to spatial-data structures and analysis, applications relevant to civil engineering, and basic skills to deal with spatial-data problems that can be solved using GIS.
Syllabus	Basic concepts including terminology, origin, and history; hardware and software; data acquisition, structure, and management; spatial analysis and modeling; and generic applications
Contents	Introduction to GIS What is GIS? GIS history Current developments of GIS Application perspectives in civil engineering GIS hardware and software Current software developments Software requirements Hardware requirements Introduction to the software used in the course Spatial data manipulation Spatial data model and input Basic data structures Spatial data transformations Output and visualization Database management Spatial analysis Basic spatial analysis functions Map analysis Analysis applications Civil engineering applications A demonstration An assignment (real engineering project)
Organization	Lectures: 3 hrs/wk Laboratory exercises: 2 hrs/wk

1. *Demonstration Teaching*: In the classroom, only one computer is used for running the GIS software with a projector and a large screen controlled by the instructor simultaneously. The purpose of this system is to show how GIS works and help students understand what the instructor teaches without any hands-on experience for students. The students need not know how to use specific GIS software because only the instructor use GIS to make them understand the theory better. But students could still learn the basic operation of the software after this live demonstration. This method is suitable to those institutions that cannot afford setting up a GIS lab, especially those in developing countries.

2. *Hands-on Interactive Learning:* In the classroom, all students can access the same GIS environment that is accessed by the instructor through multiple computers, workstations, or graphic terminals. No projector and large screen are needed because the students have their own screen. They can access the system either individually or in groups. The best grouping is two students per computer in order to encourage some in-class discussion. More than two per computer dramatically reduces the amount of time each student has for hands-on interaction based on teaching experience of two years (ESRI 1995b). In this scenario, students need know how to use the GIS software installed on their computers because they have to control the GIS by themselves.

3. *Self-Paced Learning:* This is a solution for students to learn GIS outside the classroom using open computer laboratories, special GIS laboratories, or other university networks. Students can either study how to use GIS software or aid their understanding of GIS-related courses. Knowledge of how to use GIS software is also needed to aid hands-on practice. The instructor's responsibility is to teach the basic software operation and control the students' progress.

Among the preceding methods, the last two may involve working with a computer within a university or a laboratory network. In this case, we should be aware that the networks may pose some unexpected challenges and opportunities (ESRI 1995b). Certainly, the network setting provides more access to students to gain hands-on experience and allows sharing expensive GIS facilities. However, it can slow the students' progress because the students have to learn first the basics of the network operating system such as Unix (Chow 1996). If the network setting is used, the students' prerequisite knowledge should be taken into account.

New developments on Internet and Intranet networks also provide efficient ways for educating GIS professionals. The Internet, WWW, and client/server technologies can play two important roles in GIS education in civil engineering. One is using their enormous GIS resources to provide on-line GIS tutorials and courses for GIS classroom and distance education in universities and colleges. Another is to provide accessible sites for self-improvement of professionals. The Internet can even provide a trouble-shooting aid and a discussion forum for students, such as the GIS Discussion Group at McMaster University. Students who are interested in GIS and CE professionals who want to improve their professional skills can now learn GIS by (1) taking courses at universities/colleges, (2) registering as distance education students studying through either the Internet or the conventional paper materials, and

(3) self-studying through various Internet tutorials and training courses. Table 5 lists some GIS educational sources on the Internet. These include tutorials provided by software developers and other scientists, academic courses developed by universities, and training courses developed by organizations.

INFRASTRUCTURE NEEDS

GIS education in civil engineering requires a supporting infrastructure that includes technological resources, human resources, and financial/administrative resources. In the following sections, we will discuss the technological resources (software and hardware, data sources, and laboratory setting) along with current developments.

Software and Hardware

Not all computers in a classroom need to be full-power stations. The most important criteria for hardware and software selection are that they are affordable within educational budgets and satisfy educational functionality. One of the characteristics of GIS education is the nature of the field of study: GIS is a rapidly changing field of study, which implies that many subjects change faster than a GIS tutor can keep up with (Cremers 1996). So, do the hardware and software. The basic requirements for a GIS laboratory environment should be that it aids instructors in teaching fundamental theory of GIS class and demonstrate spatial relationship to assist other classroom teaching if it will be used as a CAI supporting environment simultaneously. These requirements would be different when the laboratory is also used for research.

The annual GIS World of 1993 has reported details on over 200 different GIS software and related products, that are currently available in the GIS market (GIS World 1993). These software have different application orientations. Table 6 lists some of the currently available software that might be suitable for GIS education in civil engineering. For educational purposes, the software should have the basic GIS functions and be capable of running on a desktop platform. In selecting educational software, the following features are desirable:

- the software is friendly and has a simple and easy user interface.
- the software has tutorial materials, including software and paper documents.
- the software has a special price for educational purposes.
- the software provides sample datasets that could be used in teaching.
- the software provides multimedia interface.

Table 6 shows the desktop GIS software that can be run on the DOS or Windows platform. These PC-based software provide an affordable solution for institutions that cannot establish a comprehensive GIS laboratory and offer great potential for

Table 5　Some Internet GIS courses and tutorial resources

Course/tutorial	Homepage URL	Course contents
GIS Analysis with Arc/Info	http://boris.qub.ac.uk/ shane /arc/ARChome. Html	Introductory course on Arc/Info with lab data
Project ASSIST	Http://www.geog.le.ac. uk/assist/	Includes two introductory courses related to Arc/Info and IDRISI
Understanding GIS: Arc/Info Method	Http://ice.ucdavis.edu/ local/gis/arctutin.html	Basic concepts, functions, and commands of GIS, based on Arc/Info GIS software
GIS Tutorial	Http://www.sfrc.ufl.edu/ Laboratories/GIS/acfgis/ tutorial/acftutor.html	A tutorial on GIS, emphasizing Arc/Info and the Austin Carey Forest
Forestry 4215 (GIS course)	http://www.lakeheadu.ca /~foredwww/4215.html	Distance education version about basic knowledge of GIS and how to use Arc/Info with lab data
IDRISI Tutorial on WWW	http://www.sbg.ac.at/ geo/ idrisi/wwwtutor/ tuthome.htm	Introduction to IDRISI capabilities and ways of solving spatial problems
GIS and Remote Sensing (GRASS Tutorials)	http://liberty.uc.wlu.edu/ ~dharbor/gisrs	Introduction to GIS concepts: raster maps, DEM, overlay, digitizing, data conversion, and image processing
GIS	http://www.usgs.gov/ research/gis/title.html	A very simple introductory primer on GIS
The GIS Glossary	http://www.esri.com/ library/ glossary/ glossary.html	A GIS glossary provided by the Environmental Science Research Institute
AGI GIS Dictionary	http://www.geo.ed.ac.uk /root/agidict/	A dictionary of GIS terminology containing about 300 terms

promoting GIS education in developing countries. They also can save the time needed to learn the basics of Unix and other computer software (such as open Windows™) which are required by non-desktop systems (Chow et al. 1996). For the software listed in Table 6, the minimum requirement for the processor is 486DX. The software also need at least 8MB RAM (preferably 16 MB) and a mouse support.

Table 6   Some desktop GIS software suitable for use in civil engineering education

Software name	Minimum system requirements						Tut. Soft.	Soft. cost[a]
	Processor	RAM (MB)	HD For data	Monitor	CD-drive	Oper. system		
PC-Arc/Info	486/higher	4	-[b]	VGA	yes	Win/DOS	_	$5500 US
Atlas*GIS	386/higher	4	-	VGA		Win/DOS	_	$1600 US
ArcView	486DX	16	25MB	C-VGA	yes	Win	_	$1800 US
IDRISI for Windows	386/higher	4	-	VGA	no	Win	yes	$125 US
IDRISI for DOS	AT/(PS/2)	512k	-	E/VGA	no	DOS	yes	-
MapInfo	486DX	8	15MB	VGA	yes	Win	_	$349 US
CARIS GIS	486DX	8	60MB	C-VGA	yes	Win	yes	$99 CAN

[a] These prices may not be exact because some of them did not come from the vendors directly. The prices are for 1998.
[b] A dash means the information could not be found on available literature.

Data Sources

Data used in GIS education can be obtained from the following sources: (1) datasets included in the software package such as census tracts, ZIP codes, major roads, highways, topographic images, Digital Elevation Model data, and demographic attributes, (2) data related to spatial information, metadata, and on-line maps from the Internet, and (3) datasets established by individuals from universities for specific classroom teaching purposes.

In order to promote the development of GIS education in civil engineering, inter-agency cooperation should be encouraged so that data from different sources (e.g. universities, software vendors, and public agencies) can be exchanged. Data is the major fuel of the GIS engine. Unfortunately, it is costly to form a database even for the purpose of education, and therefore data sharing is especially important for promoting GIS education. In this regard, it is suggested that a data catalogue be established among universities for researching and retrieving datasets, where certain universities focus, for example, on transportation and others on water resources.

Laboratory Setting

Setting up an educational GIS system in the laboratory requires careful consideration. The following factors should be taken into account: (1) strong administrative/faculty support, (2) good lab design that meets the intended purpose, (3) clear, well-enforced management policies, (4) adequate funding for maintenance and upgrading, and (5) adequate staffing. Identification of the purpose is most important because it will determine the scale, site, staff, and system setting. A laboratory for the purpose of both education and research is certainly not the same as one for education only. The following are three different ways for establishing a GIS laboratory:

1. *Sharing approach for providing GIS access:* The computer center provides hardware, while the CE department provides GIS software and application-related data, models, and materials. Also, the library provides general data including map resources, photogrammetry information, and so on. An example of this setting can be found at the University of North Carolina.
2. *Integrating the GIS laboratory with existing AutoCAD laboratory using available hardware and technical resources:* The AutoCAD laboratory may have two desktop computers, two printers, one digitizer and one plotter as well as some CAD and graphics software. An example is the newly developed GIS & AutoCAD laboratory at Ningxia Institute of Technology (NIT) (Li and Shi 1995). In order to establish a GIS/AutoCAD laboratory in the Department of Civil Engineering, NIT provided financial support to buy a scanner and the department bought the GIS software.
3. *Establishing a special-purpose GIS laboratory by individual departments that can be used for education and research:* This option requires investment for the entire hardware and software, including technician training and facilities maintenance. The option is especially suitable for those departments whose faculty members have GIS-related research grants that could be partially used to establish a GIS laboratory.

To establish a GIS laboratory in either of the preceding ways, there are different hardware and software settings that depend on the type of the laboratory to be established. However, based on Table 6, the following setting choices represent the minimum requirements:

(a) *Desktop setting:*
486DX, 4-8MB RAM, and 100-200 MB hard disk space
Graphic printers, digitizer, and plotter

DOS or Windows platform
Scanner if used for research
(b) *Network setting:*
Workstations or graphic terminals connected to university/departmental networks
Graphic printer, digitizer, and plotter access
Unix or Win NT or other required platforms
Scanners access at certain locations (if used for research).

It should be noted that the preceding requirements for desktop setting represent the absolute minimum. For example, at University of New Brunswick, some 486 computers that run MapInfo, IDRISI, and CARIS for Windows are still used in teaching GIS and mapping courses. If budget is available, however, higher-performance machines should be used.

CONCLUSIONS

Geographic information systems have not been widely implemented in CE education. However, GIS has already been widely used in diverse civil engineering areas, including transportation engineering and infrastructure management. Equipped with the knowledge of GIS technology, civil engineers will better serve society in the next decade. Besides, GIS is a challenging technology with spatial analysis techniques, data management capabilities, real-time problem solutions, and diverse outputs. Integrating GIS into CE education, as a GIS course or as a CAI tool, can aid student's abilities in problem solving, critical thinking, and communications. Based on this study, the following comments regarding future developments are offered:

1. Integrating GIS concepts into CE curriculum still faces many challenges. Besides the financial and technical problems, the lack of understanding the importance of GIS education by civil engineering educators may be one of the most critical issues that deserve special attention.
2. Integrating GIS with multimedia technology for added teaching support would create a more powerful educational tool. Also, the effects that new technologies, such as telematics and the Internet, have on GIS educational need to be examined.
3. GIS software developers should pay more attention to implementing GIS in CE education. Because there are many GIS software with different industry standards in the market, it is difficult for educators to teach students how to use software. Besides, using specific software with certain application orientation may sometimes cause certain confusion because GIS is not just software. GIS software having the most common GIS-related functions is considered an acceptable educational tool.
4. GIS resources on the Internet and GIS on-line activities will benefit education. GIS on-line education should be included in the study domain of spatial information systems. This will benefit not only civil engineers, but also other professionals.

5. The CE professional societies should promote more discussion on the following aspects: (a) standards for introductory GIS courses that are transferable, (b) educational level and relevant contents of GIS courses, and (c) the need, scope, and depth of GIS education in civil engineering.
6. Concentrated effort should be devoted to integrating GIS into civil engineering education. This integration depends on several factors including (a) accreditation standards, (b) research interests of faculty members, (c) availability of laboratory facilities, (d) need for curriculum innovation, and (e) employment needs.

REFERENCES

Chang, P.C., McCuen, R.H., and Sircar, J.K. (1995). Multimedia-based instruction in engineering education: strategy. *J. Profl. Issues in Engrg. Educ. and Pract.*, 121(4), 216-219.

Chow, A., Wojnarowska, M., Finley, D., and Coleman, D. (1996). GIS for the curious: an educational tool for GIS. *White Paper #23 of Universal systems Ltd.*, Fredericton, NB, Canada.

Cremers, P. (1996) *Distance learning on GIS: facts, considerations, and questions.* Paper presented at the Symposium on GIS in Higher Education (GISHE), September 5-8, Columbia, Maryland.

ESRI (1995a). Exploring common ground: the educational promise of GIS. *An ESRI White Paper, Envir. Sys. Res. Inst., Inc.*, Redlands, CA.

ESRI (1995b). GIS in K-12 education. *An ESRI White Paper, Envir. Sys Res. Inst., Inc.*, Redlands, CA.

GIS World, Inc. (1993). *GIS sourcebook.* Fort Collins, Colo.

Horak, J., Jancik, P., and Rapant, P. (1994). Proposed complex system of GIS education at Faculty of Mining and Metallurgy. *EGIS Foundation*, Internet: http://www.odyssey.maine.edu/gisweb/spatdb/egis/

Li, S. and Shi, Y. (1995). *Proposal for establishing GIS/AutoCAD labortory at institute level.* Tech. Report, Dept. of Civ. Engrg., Ningxia Inst. of Tech., Ningxia, China.

Lourie, I.K., (1994). GIS education in Russian universities: modern state and prospects. *EGIS Foundation,* Internet: http://www.odyssey.maine.edu/gisweb/spatdb/egis/

Committee on National Standards for Applied Science and Engineering Technologists. (1994). *National Standard - Civil Technologies.* First Edition.

Piscedda, S. (1994). How to set up a GIS course: the contribution of Formez through an Italian experience. *EGIS Foundation,* Internet: http://www.odyssey.maine.edu/gisweb/spatdb/egis/

CHAPTER 15

Looking Ahead

Yupo Chan and Said Easa

When we started this project, we underestimated the enormity of the field of GIS, both in terms of its scope and its implications for urban planing and development. The enthusiasms of our contributing authors and their convincing messages have demonstrated the great potential of this information technology well beyond our expectations. Hopefully, this book will have some impact on the practice of civil engineers, planners, managers, and public officials immediately upon its publication and in the years to come. While people may be initially attracted to GIS because of its novelty, they soon find out that it is not just 'trendy,' but essential for us to embrace this new way of organizing and analyzing spatial data. In fact, there is no other alternative!

In our opinion, the fundamental concepts behind geographic information system technology have reached a certain level of maturity and will not likely change significantly in the near future. Instead, most developments will focus on new computational environments and in applying the technology to new and challenging urban planning and development problems. At present, only a few software focuses on *integrated* applications that make a city or a region function successfully, and much is still to be done. Nonetheless, a number of hurdles related to the GIS technology, particularly database management, remains to be surmounted.

TECHNOLOGY ISSUES

In recent years, research in GIS has started to deal with some of the complexities associated with realistic spatial data. However, this flurry of activities on methods and techniques has not been accompanied by a wide dissemination into practice. To a large extent this is due to the lack of readily available software that incorporate the required spatial experiments or tests. Currently, few of the popular statistical or econometric packages for mainframes, workstations, or personal computers include any techniques for spatial analysis. The same holds to a large extent for the commercial GISs, which have seen an explosive growth in the last few years. It should be worrisome to regional scientists that the implementations of this new technology, which according to some has the potential to evolve the ideal models of spatial information, lack features to carry out all but the most rudimentary forms of spatial analysis. Hopefully, this may change over time, as major vendors such as Intergraph and Oracle recognize that up to 85 percent of all data have spatial components, and there is a market for such routines.

One of the difficulties associated with the implementation of basic GIS routines in spatial analysis is the associated computational burden. In spite of the speed of today's computational machinery, the data resolution has become increasingly more detailed for many applications and the analytical models also become more complex. In the words of Hodgson et al. (1995), we continue to create the need for faster processors to handle larger amounts of model-generated data as well as more voluminous raw-data collections. An integrated approach combining efficient algorithms (including heuristics) with the power of parallel-processing machines can be a way to overcome this problem. This harnesses the potentiality of both advances, providing us with a way to catch up with the increasing computational burden.

Thus, two opposite approaches are evident in spatial-data analysis. One is model-driven and lets spatial theory determine which specifications need to be empirically validated, as in the case of spatial econometrics. The other is data-driven and is geared toward elucidating theory from the data, as seen in recent developments in spatial statistics. Similar techniques have been developed in both approaches, with very little cross-reference between the two. Besides the loss in efficiency from parallel developments, it should be clear that the two approaches are not mutually exclusive and competitive, but complementary. With the recent gain in power and popularity of GIS, and a growing acceptance of the viewpoint of 'letting the data speak for themselves' it becomes feasible and it is crucial to keep an important role for spatial theory in spatial-data analysis.

At the same time, most data used with quantitative methods in urban planning and development are notoriously bad: the scope of available information is limited, it is often not collected in a consistent fashion, and is only loosely connected to the concepts that underlie spatial theory. Fortunately, recent advances in remote sensing have begun to address some of these problems. Ideally, a GIS with spatial data available at a disaggregate level will allow the determination of the geographic level-of-specificity to become endogenous to the analysis itself. This will hopefully mitigate the bias due to the size and configuration of spatial units such as census tracts or traffic analysis zones.

The data model implicit in a GIS is the `discretization' of geographical reality necessitated by the nature of computing devices. Commercial GIS can be classified as following either a raster or vector-data format. The raster or vector structure defines the spatial-unit-of-observation that can be used in spatial analysis. In the former, the unit is the grid and all points within the grid are assumed to take on the same value. This is an implicit form of spatial sampling. Clearly, if the grid does not exactly correspond to the spatial arrangement of values in the underlying process there will be an inherent tendency of spatial dependence. In other words, geographic neighbors show similarities between them, not driven by facts, but by the process. Similarly, if the scale of the grid cell has an imperfect match with the scale of the process studied, various types of mis-specification may result, often referred to as ecological fallacy or the modifiable-areal-unit problem. For example, a model for choosing residential location would be based not on commuting or shopping patterns, but on spurious correlation between misclassified data points.

When a vector structure is used, the choice of the points, lines, and polygons that will be presented, their spatial resolution, and spatial arrangement are also an implicit form of spatial sampling. Similar to the raster approach, homogeneity is assumed within the point, line, or areal-unit of observation. For the latter in particular, this may only be a crude approximation and spatial dependence as well as scale problems are likely to be present. Raster format may be in its ascendancy due to the advances in remote sensing and digital computing technologies. Renewed interest in Voronoi diagrams, however, provides an equally invigorated interest in vector-based data storage. Thus the same analysis issues would face vector-based GIS as they haunted raster-based systems.

In the discussions above, we have introduced remote sensing as a viable source of real-time information (particularly after processing). To the extent that aerial photography is one kind of remote sensing, the concept is not new. The large number of channels available in today's satellites, however, makes available information that the bare eye cannot see, affording infrared and heat-sensing signals that are essential in urban planning and development applications. Geosynchronized satellite constellations also provide real-time raster images to a fine level of resolution, not to say the GPS that is used extensively for navigation (US Department of Transportation 1995). These technologies have far-reaching implications in planning and development.

DATABASE ISSUES

The data-sampling process structures the database and precedes any sampling the engineer/planner himself/herself may want to carry out. It is often dictated by administrative or policy concerns that may be founded on `accepted' theoretical concepts of the time. Examples are the delineation of administrative regions (such as school districts or census tracts) that pre-determine the collection of many socioeconomic data. In a sense then, although spatial analysis may be exploratory, the data that are available and the way in which they are collected and arranged are often constrained by the accepted theoretical knowledge of the time and its implications for spatial resolution. Digitization into a raster format may help in data standardization. Concomitant with it, however, is the expense of conversion and storage. It may also be a simplistic way of addressing this problem. Meanwhile, extensive efforts have been carried out recently on developing a spatial-data infrastructure (National Research Council 1994) and spatial database transfer-standards (Mollering 1997). The huge amount of work involved testifies to the complexity and importance of this subject.

Other database issues are also emerging. First, interoperability, that signifies the ability of multi-vendors and multi-platform systems to interact with each other by changing data and functions, is an important issue in urban planning and development applications because of the complexity of the data, operations, and analysis used. In such systems (called *open systems*) the user is not bound to a particular vendor to extend the system functionality by adding new components or linking with other systems. There is a growing demand for such systems, as exemplified by the *Open GIS Consortium* which is an independent partnership of universities and industry (Burrough and McDonnell 1998). Some advances in open systems are presented in Chapter 4.

Second, Internet-based GIS, that provides users access to geographic data across the Internet, is becoming more useful for third-party GIS applications. While Internet-based GIS is still in its infancy, practical projects already exist that can help in such applications as planning and management of lifeline infrastructure systems. Finally, knowledge discovery in the databases (KDD) is a challenge for researchers to find useful information in rapidly growing databases, such as patterns and trends. Future research in this area will in the short-term focus on developing efficient KDD algorithms and in the long term on developing query languages for KDD, query processing techniques, and interoperating with application program interfaces. For more details on emerging database issues, the reader is referred to Adam and Gangopadhyay (1997).

Geographic information systems allow the merging of data from diverse sources, from remote sensing to survey and interview data. Modern data-processing capabilities such as relational database and object-oriented programming do not only facilitate data fusion, but also greatly streamline modeling applications, including spatial analysis. To the extent that spatial relationship is the basic building block for urban planning and development, GIS becomes an integral part of today's analysis toolkit. It has been shown in this book that a very desirable focus of GIS is problem solving. With the convenience of electronic data-transfer, GIS is also a *Global* information system, affording truly distributed decision-making to take place anywhere in the world through real-time sharing of information.

REFERENCES

Adam, N.R. and Gangopadhyay, A. (1997). *Database issues in geographic information systems*. Kluwer Academic Publishers, Boston, MA.

Burrough, P.A. and McDonnell, R.A. (1998). Principles of geographical information systems. Oxford University Press, New York, N.Y.

Hodgson, M.E., Cheng, Y., Coleman, P., and Durfee, R. (1995). "Computational GIS burdens: solution with heuristic algorithms and parallel processing." *Geo Info System*, April.

Mollering, H. (ed.). (1997). *Spatial database transfer standards 2: characteristics for assessing standard and full descriptions of the national and international standard in the world*. International Cartographic Association and Pergamon Press.

National Research Council. (1994). *Promoting national spatial data infrastructure through partnerships*. Commission on Geosciences, Environment, and Resources, National Academy Press, Washington, D.C.

U.S. Department of Transportation. (1995b). "Coast Guard announces expected differential global positioning system operational startup." *News*, CG 45-95, December 22.

Selected Internet Resources

This appendix presents a wide variety of useful Internet resources that are grouped under the following categories: GIS resources, GPS and remote sensing resources, online spatial data resources, software and applications, Internet GIS/Map server and interface, GIS online glossary, and other. Both the Internet site address (URL) and a brief description of the site are presented. In some cases the sites are themselves lists of hundreds of other GIS and related sites.

GIS RESOURCES

HTM Great GIS Net Site
http://www.hdm.com/gis3.htm
This comprehensive site includes many Internet resources on GIS books, GIS conferences, breaking GIS news, classic GIS sites, the best web resource lists, online GIS, GIS data and software library, GIS jobs, worldwide government agencies, GIS software companies, and GIS/mapping companies. There are also gateways to other GIS WWW resources.

GISLink Resources
http://www.gislinx.com/
GISLink is a categorized list of GIS sites. There are 26 categories such as software, hardware, publications, GIS in Canada, GIS events, and Java/GIS resources and data. A search engine is provided to allow users to search the site.

GIS Resources Dictionary
http://www.geojobsource.com/resource.htm
This site, provided by the Geo Job Resource, has three major categories, including academe, professional associations, and other resources. From here, you may find most academic organizations related to GIS, including university departments and associations.

GPS AND REMOTE SENSING RESOURCES

GPS Tutorial
http://www.trimble.com/gps/fsections/aa_f0.htm
A quick tutorial on the basic principles behind GPS technology by Trimble Navigation Ltd.

Paul Tarr's GPS WWW Resources List
http://www.inmet.com/~pwt/gps_gen.html

This site provides a table of contents and links to introductory GPS material, general GPS and navigation related information, non-commercial and commercial web sites, related activities and GPS business inquires.

Remote Sensing Resources
http://observe.ivv.nasa.gov/nasa/education/reference/main.html
A lot of remote sensing basic stuff includes most aspects. There are also links to many online tutorials related to remote sensing.

ONLINE SPATIAL DATA RESOURCES

National Geospatial Data Clearinghouse
http://fgdclearhs.er.usgs.gov/
This site enables access to spatial data with accompanying metadata by searching through a single interface based on their descriptions, or 'metadata.' Several entry points have been established at this federal geospatial clearinghouse website, including EPA and USGS.

U.S. EPA Environ-Facts Warehouse
http://www.epa.gov/enviro/html/ef_home.html
This site provides environmental data such as those used to pinpoint toxic releases in EPA's inventory for a region by site, zip code, county, basin, or facility information. Also, information on waste discharge permits, grants, air quality, and EPA's spatial data library can all be accessed from this site.

FEMA Digital Flood Data On-line
http://www.fema.gov/
The FEMA Map Service Centre allows users to download spatial Q3 Flood Data directly or place an order for delivery on a CD-ROM. By July 1997, 800 counties acre were made available.

USGS DLGs, DRGs, and DCWs
http://www.usgs.gov/pubprod/
Along with digital data of all types and formats, this USCG site includes water resources distributed data, national stream quality, accounting network stations, HUCs, LULCs, reservoirs, annual U.S. runoff, stream flow data, CERCLA and RCRA sites info, landfill locations, and quads. A full explanation of GIS and useful GIS sites are found at: http://internet.er.usgs.gov/research/gis/title.html

U.S. Census Bureau Homepage and GIS Gateway
http://tiger.census.gov/
Connections to many GIS sites are included on the Gateway page and access to TIGER files and other spatial information is provided, including county and state boundaries.

U.S. State Webpages
Almost all states are in the process of developing sites to access information and spatial data. Below are a few of these sites:

- *Florida Maps.*
  http://www.floridamaps.com/ & http:/www.floridamaps.com/fdep.html
  Data developed by Florida Department of Environmental Regulation (FDEP).
- *Colorado's Natural Resources Online.*
  http://www.dnr.state.co.us/index.asp
- *Montana Natural Resource Information System.*
  http://nris.msl.mt.gov/gis/mtmaps.html
  Statewide coverage and metadata in several formats
- *Iowa Natural Resources.* GIS
  http://www.igsb.uiowa.edu/nrgis/gishome.htm
  Includes political boundaries and physical and environmental geography.

Guide to Online/Free U.S. Geospatial Data
http://www.cast.uark.edu/local/hunt/index.html
This site is a good starting point to track down sources of digital geospatial data and attributes related to the U.S., including Geospatial Data Servers for different states and cities, and different data formats. The site is organized by different categories, levels, application areas, and data providers.

GIS Data Resources Directories
http://sunsite.berkeley.edu/GIS/gisdata.html
Data is provided that may be useful in GIS project development, including (1) resource directories for GIS, remote sensing, and related tabular data resources, (2) federal agencies, NSDI nodes, and data search services, (3) state, regional, and local data holdings, and (4) data from other countries, like Australia and Romania.

SOFTWARE AND APPLICATIONS

Environmental Systems Research Institute (ESRI)
http://www.esri.com/
The Business Partner Authorised Developers pages list 3rd party software developers for use with ESRI GIS products, including environmental, hazardous waste, and water/wastewater products. Abstracts from the yearly ESRI conference and a DataHound for spatial data hunting are also included here.

GIS and Model Integration
http://www.compass.ie/gis/models.html
This site lists many models that have been integrated with GIS along with Internet links to their developers.

Java/GIS Resource Listing
http://www.gislinx.com/java.html

This site provides Internet links to companies, source code providers, research projects, and application sites that are related to Java and GIS, especially using Java technology to develop online GIS applications.

GeoJava Corner
http://www.ggrweb.com/geojava/index.html
This site includes Java spatial data viewer and other utilities for Internet GIS applications development, especially many links to the online Java resources including language and development tools.

INTERNET GIS/MAP SERVER AND INTERFACE

Virtual Boston
http://www.pmg.lcs.mit.edu/~ng/Map/
This site includes a simple example of what Java can do with maps. Java applet is used to provide a map interface that allows the user to surf the Boston area.

MapBlast
http://www.mapblast.com/mblast/index.mb
With the exception of online map service, this site provides other information such as step-by-step driving directions and maps as well as news, weather, and events of interests. It also allows you to find local stories and businesses.

MapQuest
http://www.mapquest.com/
This site provides online maps by looking for an address, town, or zip code, driving directions from door-to-door or city-to-city, and other information such as travel guide.

Etak Digital Map
http://www.etakguide.com/
This site is an online digital map provider. EtakGuide will find a location anywhere in the United States and centre a map on the spot you have chosen. The site allows for zoom in/out and pan operations.

GIS ONLINE GLOSSARY

AGI Online GIS Dictionary
http://www.geo.ed.ac.uk/agidict/welcome.html
This site provides online GIS dictionary that includes definitions for 980 terms compiled from a variety of sources either related directly to GIS or encountered by GIS users in their work. 52 diagrams also supplement the dictionary.

Dictionary of Abbreviations and Acronyms for GIS/RS/Cartography
http://www.lib.berkeley.edu/EART/abbrev.html

This is a dictionary of abbreviations and acronyms in GIS, cartography, and remote sensing. You can use the FIND function in your web browser to search a specific abbreviation or acronym, or you can browser the alphabetical list.

ESRI GIS Glossary
http://www.esri.com/library/glossary/glossary.html
This glossary of GIS terms, maintained by ESRI, will help familiarize you with terms associated with GIS and ESRI software. You may also download this glossary as a PDF file.

OTHER USEFUL SITES

URISA
http://www.urisa.org/
This site has a lot of information related to the use of GIS and AM/FM in urban planning.

The United Nations Environment Program
http://www.grida.no/
This site provides many worldwide maps and GIS data and documents. There are also gateways to the environmental websites in different countries.

GIS World Magazine
http:/www.geoplace.com/print/gw/index.html
This site is an excellent source for the fastest trends in GIS. Many important GIS publications are published in this site by GIS World Books, Inc.

GPS World
http://www.gpsworld.com/
GPS World has been tracking the growth of this new multibillion-dollar industry, composed of users, researchers, system developers, and manufacturers. This online magazine will help you enter the world of GPS, GLONASS, and satellite positioning and you will discover an expanding universe of new applications proliferating with the rich diversity of human imagination.

Spatial Odyssey
http://wwwsgi.ursus.maine.edu/biblio/home.html
This bibliography (from 1991 onward) lists conference proceedings and various compendiums of interest to GIS researchers. Major conferences include ACSM/ASPRS, AM/FM, AURISA, Canadian Conference on GIS (Ottawa, ON), EGIS, GIS/LIS, GIS Symposium (Vancouver, BC), GIS for Transportation, and Symposium on Geographic Information Systems and Water Resources. The full texts of papers from some conference proceedings are provided.

Acronyms

2-D	Two-Dimensional
2.5-D	2.5-Dimensional
3-D	Three-Dimensional

A

ACN	Automated Collision Notification
ACSM	American Congress of Surveying and Mapping
ADT	Abstract Data Type
AGI	Association for Geographic Information (UK)
AHP	Analytical Hierarchy Method
AHS	Automated Highway System
AI	Artificial Intelligence
AM/FM	Automated Mapping and Facilities Management
AM/FM/GIS	AM/FM integrated with GIS
AMR	Automatic Meter Reading
AR/GIS	Active Response GIS
ASPRS	American Society of Photogrammetry and Remote Sensing
AURISA	Australian Urban and Regional Information Systems Association

B

BTS	Bureau of Transportation Statistics

C

CAD	Computer Aided Design (or Drafting)
CAI	Computer-Aided Instruction
CARIS	Computer-Aided Resource Information Systems (a trade mark of Universal Systems Ltd. in Canada)
CARL	Central Attributed Raster Line
CCT	Computer Compatible Tape
CDA	Community Development Agency
CD ROM	Compact Disc Read Only Memory
CE	Civil Engineering
CERCLA	Comprehensive Environmental Response, Compensation, and Liability Act
CO	Carbon Monoxide
COBRA	Common Object Request Broker Architecture
COG	Council of Governments
COGO	Coordinate Geometry
COM	Component Object Model

D

DA	Distribution Automation
DBMS	Database Management System
DCW	Digital Chart of the World
DDE	Dynamic Data Exchange
DEI	Data Extraction Interface
DEM	Digital Elevation Model
DGI	Distributed Geographic Information (via the INTERNET)
DLG	Digital Line Graphs
DOE	United States Department of Energy
DOS	Disk Operating System
DOT	Department of Transportation
DRAM	Disaggregate Residential Allocation Model
DRCOG	Denver Regional Council of Governments
DRG	Digital Raster Graphics
DTD	Document Type Definition
DTM	Digital Terrain Model
EMPAL	Employment Allocation Model

E

EIS	Environmental Impact Statement
EGIS	European Geographical Information System
EPA	Environmental Protection Agency (U.S.)
EPAD	Edge Probability Attributed Data
ERTS	Earth Resources Technology Satellites
ES	Expert Systems
ESRI	Environmental Systems Research Institute (a GIS and Mapping software company in the U.S.)

F

FDEP	Florida Department of Environmental Regulation
FEMA	Federal Emergency Management Agency
FGDC	Federal Geographic Data Committee
FIS	Flood Insurance Study
FME	Feature Manipulation Engine

G

GIA	Geographic Information Analysis
GIS	Geographic Information Systems
GIS-BARS	Geographic Information System Based Accident Records System
GIS-T	Geographic Information System for Transportation
GLONASS	Global Navigation Satellite Systems
GPS	Global Positioning System
GTS	Ground Transportation Subcommittee of the FGDC

| GUI | Graphical User Interface |

H

HC	Hydrocarbons
HCD	Housing Conservation District
HLRW	High Level Radioactive Wastes
HPMS	Highway Performance Monitoring System
HTM	Harvard Design and Mapping Company, Inc
HUC	Hydrologic Unit Code

I

IFOV	Instantaneous Field of View
IMMPS	Intelligent Multimedia Presentation Systems
IMS	Intersection Modeling System
ISTEA	Intermodal Surface Transportation Efficiency Act
ISWM	Integrated Solid Wastes Management
IT	Information Technology
ITE	Institute of Transportation Engineers
ITS	Intelligent Transportation Systems
IVHS	Intelligent Vehicle Highway Systems

L

LAN	Local Area Network
LIS	Land Information Systems
LPM	Lines Per Millimeter
LRA	Land Redevelopment Authority
LRM	Location Reference Method
LRS	Linear Referencing Systems
LRT	Light Rail Stations
LULC	Land Use/Land Cover

M

MCDA	Multi-Criteria Decision Analysis
MEASURE	Mobile Emission Assessment System for Urban and Regional Evaluation (EPA)
MIS	Management Information System
MP	Mile Point
MPO	Metropolitan Planning Organization
MSS	Multi Spectral Scanner
MTU	Master Terminal Unit
MUTCD	Manual of Uniform Traffic Control Devices
MWRA	Massachusetts Water Resources Authority

N

| NAAQS | National Ambient Air Quality Standards |

NCHRP	National Highway Cooperative Research Program
NIJ	National Institute of Justice
NIMBY	"Not In My Back Yard!"
NOAA	National Organization of Air and Aeronautics:
Nox	Nitrogen Oxides
NPDES	National Pollution Discharge Elimination System
NSDI	National Spatial Data Infrastructure
NSF	National Science Foundation

O

OCX	OLE/ActiveX custom control
ODBC	Open Database Connectivity
OLE	Object Linking and Embedding
ORDBMS	Object Relational Database Management System

P

PDF	Portable Data Format
PGML	Precision Graphics Markup Language
PPI	Plan Position Indicator
PROMETHEE	Preference Ranking Organization Method for Enrichment Evaluation

R

RAM	Random Access Memory
RAAM	Regional Activity Allocation Model
RADAR	Radio Detection and Ranging
RBCA	Risk-Based Corrective Action
RBV	Return Beam Videocon
RCRA	Resource Conservation and Recovery Act
RDBMS	Relational Database Management System
RMS	River Modeling System (BOSS International)
RS	Remote Sensing
RTU	Remote Terminal Unit

S

SAR	Synthetic Aperture Radar
SAWC	South Australian Water Corporation
SCADA	Supervisory Control and Data Acquisition
SDE	Spatial Data Engine
SDO	Spatial Data Option
SDSS	Spatial Decision Support System
SDTS	Spatial Data Transfer Standard
SLAR	Side Looking Airborne Radar
SNF	Spent Nuclear Fuels
SNR	Signal to Noise Ratio

SPOT	Satellite Pour l'Observation de la Terre
SQL	Structured Query Language
SR	State Route
SVAD	Single Value Attributed Data
SWMM	Stormwater Management Model (EPA)

T

TAZ	Transportation (Traffic) Analysis Zone
TIGER	Topologically Integrated Geographic Encoding and Referencing
TIN	Triangulated Irregular Network
TIP	Transportation Improvement Plan
TGIS	Temporal GIS
TM	Thematic Mapper
TMIP	Transportation Model Improvement Program
TNP	Transportation Network Profile
TSDD	Transportation Spatial Data Dictionary (draft)

U

UAM	Urban Airshed Model
URISA	Urban and Regional Information Systems Association
URL	Uniform Resource Locators
USDOT	United States Department of Transportation
USGS	United States Geological Survey
UTM	Universal Transverse Mercator projection

V

| VML | Vector Markup Language |
| VMT | Vehicle Miles Traveled |

W

| WAN | Wide Area Network |
| WWW | World Wide Web |

X

| XML | eXtensible Markup (meta) Language |

Z

| ZAI | Zonal Attractiveness Index |